Advanced Coal, Petroleum and Nature Gas Exploration Technology

Advanced Coal, Petroleum and Nature Gas Exploration Technology

Editors

Gan Feng
Qingxiang Meng
Fei Wu
Gan Li

MDPI • Basel • Beijing • Wuhan • Barcelona • Belgrade • Manchester • Tokyo • Cluj • Tianjin

Editors
Gan Feng
Sichuan University
China

Qingxiang Meng
Hohai University
China

Fei Wu
Chongqing University
China

Gan Li
Ningbo University
China

Editorial Office
MDPI
St. Alban-Anlage 66
4052 Basel, Switzerland

This is a reprint of articles from the Topical Collection published online in the open access journal *Energies* (ISSN 1996-1073) (available at: https://www.mdpi.com/journal/energies/topical_collections/exploration_technology).

For citation purposes, cite each article independently as indicated on the article page online and as indicated below:

LastName, A.A.; LastName, B.B.; LastName, C.C. Article Title. *Journal Name* **Year**, *Volume Number*, Page Range.

ISBN 978-3-0365-6682-5 (Hbk)
ISBN 978-3-0365-6683-2 (PDF)

Contents

Editorial

Advanced Coal, Petroleum, and Natural Gas Exploration Technology

Gan Feng [1], Hongqiang Xie [1,*], Qingxiang Meng [2], Fei Wu [3] and Gan Li [4]

[1] State Key Laboratory of Hydraulics and Mountain River Engineering, College of Water Resource & Hydropower, Sichuan University, Chengdu 610065, China
[2] Research Institute of Geotechnical Engineering, Hohai University, Nanjing 210098, China
[3] State Key Laboratory of Coal Mine Disaster Dynamics and Control, Chongqing University, Chongqing 400044, China
[4] Department of Civil Engineering, Ningbo University, Ningbo 315211, China
* Correspondence: alex_xhq@scu.edu.cn

Citation: Feng, G.; Xie, H.; Meng, Q.; Wu, F.; Li, G. Advanced Coal, Petroleum, and Natural Gas Exploration Technology. *Energies* **2022**, *15*, 8976. https://doi.org/10.3390/en15238976

Received: 14 September 2022
Accepted: 20 September 2022
Published: 28 November 2022

Publisher's Note: MDPI stays neutral with regard to jurisdictional claims in published maps and institutional affiliations.

Oil, coal, and natural gas are traditional fossil energy sources and the main components of primary energy consumption globally. They will also remain the world's major energy sources for a significant amount of time to come. The energy information administration's 2019 energy report predicts that global energy consumption will continue to grow until 2050 [1]. Meanwhile, however, increasingly more attention is being paid to protecting the ecological environment. On the one hand, there is a movement to gradually seek, develop, and utilize renewable green energy. From 2020 to 2021, global renewable energy consumption increased by 15%. On the other hand, people are also constantly exploring high-efficiency fossil energy exploitation technologies that are environmentally friendly and reduce the negative environmental impacts of these energy sources while meeting the human demand for energy.

According to the relevant research data of some scholars, the world's recoverable oil reserves are about 95.6×10^{10} t, of which unconventional oil resources account for about 44%. The estimated recoverable natural gas reserves are 783.8×10^{12} m^3, of which about 25% comprise unconventional natural gas resources. It is estimated that global unconventional oil production will exceed 10×10^8 t, accounting for about 20% of the world's total crude oil supply, and that global unconventional natural gas production will increase to 42% of the world's total natural gas production [2].

Traditional oil recovery methods, i.e., primary oil recovery and secondary oil recovery, can help make the recovery rate of original oil reach 20~40% [3,4]. The oil recovery efficiency must be continuously improved to ensure that people have sufficient use of oil resources [5]. Therefore, researchers have been promoting research into advanced oil extraction technology. For example, there has been progressed in the application of ultrasound technology for demulsification and enhanced oil recovery (EOR), large capacity hydraulic fracturing, and special horizontal wells for shale oil and gas production [6]. The fracturing of platform wells is key to the efficient development of unconventional oil and gas, and it has been widely used in oil fields. Field practice shows that different fracturing methods lead to different stimulation effects [7]. The permeability and porosity of unconventional reservoirs are usually relatively low, and hydraulic fracturing is often used to improve the flow performance of the formation, so as to achieve efficient and economic development [8]. Stress differences, rock brittleness, natural fractures, bedding, cluster spacing, fracturing method, fracturing fluid volume, displacement, and viscosity have important influences on fracture network complexity (FNC) [7]. Nanofluids and nanomaterials are great prospects for developing new technologies in the petroleum sector [9]. Therefore, as new technologies for oil and gas exploitation are constantly put forward and improved, it will promote more efficient, safe, and economic development of oil and gas. It is also very important for researchers to continue to pay attention to and research new technologies for oil and gas development, and pay attention to the application of these new technologies in engineering.

The world is rich in shale oil resources. As the United States has made excellent research achievements in shale oil exploration and exploitation, shale oil has become a major contributor to the growth of crude oil production [10]. The successful exploration and economic development of marine shale oil in the United States has benefitted from the continuous improvement of basic geological knowledge [11]. At present, high-yield shale oil wells in the United States are mostly distributed in the Williston Basin, the Permian Basin, and the Gulf Coast Basin [12]. Shale oil exploration in China is mainly concentrated in continental basins. A Research Report points out that medium and high maturity shale oil is the key area for China's strategic shale oil breakthrough, and strengthen the exploration and development of shale oil and gas [13].

Unconventional natural gas resources mainly include tight sandstone gas, coalbed methane, and shale gas. Unconventional gas reservoirs usually refer to low permeability gas reservoirs that do not contain or contain a small amount of associated fluid in the produced natural gas. In general, large-scale hydraulic fracturing, horizontal wells, multi-branch wells, and other technologies are used to exploit these kinds of resources economically and effectively. Natural gas development technology includes fractured gas reservoir development technology, condensate gas reservoir cyclic injection development technology, gas lift drainage technology, machine pumped drainage technology, deep carbonate gas reservoir development technology, and deep tight sandstone gas reservoir group development technology, among others. The development technology of these resources is of great significance for the exploitation of unconventional natural gas resources under various geological causes and occurrence conditions. In the future, it is not only necessary to deeply study these technologies, but also to explore new technologies for unconventional natural gas development in various complex environments, and the actual application effect of these new technologies should also be verified by the actual project site.

As a clean fossil fuel with huge reserves, natural gas hydrate (NGH) is widely considered to be an important alternative energy source for the future. NGHs are a kind of high-concentration natural gas and exist as a crystalline compound formed under low temperature and high pressure [14,15]. They can burn directly, and so they are also called "combustible ice". NGH mainly occurs in the seabed and permafrost, where the high-pressure and low-temperature conditions required for its stability can be met. NGH has a significant impact on a series of scientific issues, such as the global carbon cycle, climate change, and geological disasters [16–18]. Several methods have been proposed for NGH exploitation [19]. Some studies have shown that depressurization may be a relatively economical and effective production method for NGHs [20]. Thermal stimulation is a very common method to improve the decomposition efficiency of natural gas hydrate. The technology of replacing CH_4 with carbon dioxide can not only produce methane but also store carbon dioxide [21]. CO_2 is injected into the deposition layer of NGHs to replace CH_4 in the hydrate and store CO_2 [22]. Since CO_2 hydrate is more stable than CH_4 hydrate at the same temperature and pressure, the CO_2 replacement method used for NGH production can maintain the mechanical stability of deposits [23]. The natural gas hydrate resource is also recognized globally as having the potential to replace the traditional fossil energy. Therefore, the research on its development technology is still the focus of energy workers. Particularly for the continuous improvement of traditional development technology and the exploration of new technology, there is a lot of research space.

Coal is a common fossil fuel and a valuable energy resource. Coal has been widely used around the world as a fuel source for transportation, industry, manufacturing, and power generation [24]. Due to the impact of COVID-19 and other factors, the world economy has slowed down. Compared with the past, the growth of global coal demand has slowed down significantly, and global coal consumption will peak in the mid-2020s [25,26]. Coal mining methods include surface mining and underground mining. With continual coal mining has come advances in coal mining technology, which has gradually been innovated upon and improved. Coal mining also requires mining equipment to adapt to the complex technological and geological changes encountered in the mining process. Lvu et al. proposed a new fully mechanized top coal caving mining method of mining the middle layer first for an extra thick coal seam above 20 m [27]. US longwall mining has employed multiple entry development using a continuous mining system for panel development and mining in retreat. In Europe, the single arched entry development and advancing system has been employed [28]. However, longwall caving mining (LCM) can lead to many environmental problems [29].

With the progress of human society, coal mining is under increasing ecological and environmental protection pressure. Coal mining and CBM mining may cause changes in surface topography, groundwater level and connectivity, soil profile, vegetation coverage, etc. [24]. This requires coal mining activities to be performed in a way that minimizes its negative impact on the environment. We should also pay attention to the restoration of the ecological environment after coal mining.

With the development of global internet technology and informatization, coal mining has become more intelligent and new advances have been made and applied in some coal mines. Due to complex mining and geological conditions, various sensors and their software control algorithms used to detect unexpected underground events must be identified, developed, and tested successfully in advance to achieve full automation [28]. Rashed et al. proposed a novel framework for assessing and prioritizing smart mining strategies by integrating the Z-number theory and the fuzzy-VIKOR technique [30]. The modern longwall mining method adopted in Germany and the United Kingdom is a true continuous mining system with great potential to rapidly increase production to meet national energy needs [28]. China has 34 coal mines (under construction, reconstruction and expansion) of 10 million-ton and 47 intelligent unmanned coal mining faces [31]. China's coal mines have the conditions required for the application of artificial intelligence, the internet, the Internet of Things (IoT), big data, and other technologies to achieve enhanced coal mining production [32]. As mines constitute an important part of a country's economy with a considerable impact on socioeconomic development, smartening mining activities should be done to increase efficiency [30]. Therefore, intelligent and ecological coal mining and environmental remediation have become the forefront of coal mining research.

Energy is an important material basis for promoting the continuous progress of human society. Scientific researchers and engineering practitioners in the global energy field are constantly upgrading and developing advanced energy mining technologies. We cordially invite you to submit your high-quality manuscripts to the Special Issue of the journal Energies on advanced coal, petroleum, and natural gas exploration technology. The keywords for this Special Issue include (but are not limited to):

➢ Unconventional resources;
➢ Reservoir characterization;
➢ Enhanced oil recovery;
➢ Multiphase flow;
➢ Numerical simulation;
➢ Energy efficiency;
➢ Petroleum geology;
➢ Hydraulic fracturing.

This Special Issue will cover a broad range of topics concerning coal, petroleum, natural gas, coalbed methane, shale oil, combustible ice, and other fossil energies, as well as exploration, reservoir characterization, machine learning applications, well logging, and geological aspects. This issue will approach these topics from both an exploration and a production standpoint. We invite papers on technical development, basic theories, reviews, experiment, and engineering case studies in the field of energy development, as well as research papers in different disciplines that are related to energy themes.

Author Contributions: G.F., Q.M., F.W. and G.L. contributed equally to the design, implementation, and the delivery of the special issue. Conceptualization, G.F. and H.X.; writing-original draft preparation, H.X., Q.M., F.W.; writing-review and editing, G.F., H.X. and G.L. All authors have read and agreed to the published version of the manuscript.

Funding: This research received no external funding.

Conflicts of Interest: The authors declare no conflict of interest.

References

1. U.S. Energy Information Administration. *Annual Energy Outlook 2019 with Projections to 2050*; U.S. Energy Information Administration: Washington, DC, USA, 2019.
2. Jia, C.; Peng, X.; Song, Y. The mechanism of unconventional hydrocarbon formation: Hydrocarbon self-sealing and intermolecular forces. *Pet. Explor. Dev.* **2021**, *48*, 507–526. [CrossRef]
3. Bera, A.; Mandal, A. Microemulsions: A novel approach to enhanced oil recovery: A review. *J. Pet. Explor. Prod. Technol.* **2015**, *5*, 255–268. [CrossRef]
4. Hirasaki, G.J.; Miller, C.A.; Puerto, M. Recent advances in surfactant EOR. *SPE J.* **2011**, *16*, 889–907. [CrossRef]
5. Ayoub, E.; Cologon, P.; Marsiglia, M.; Salmon, J.B.; Dalmazzone, C. Development of online separation and surfactant quantification in effluents from an enhanced oil recovery (EOR) experiment. *J. Pet. Sci. Eng.* **2022**, *208*, 109696. [CrossRef]
6. Adeyemi, I.; Meribout, M.; Khezzar, L. Recent developments, challenges, and prospects of ultrasound-assisted oil technologies. *Ultrason. Sonochem.* **2022**, *82*, 105902. [CrossRef]
7. Meng, H.; Wang, X.; Ge, H.; Chen, L.; Yao, Y.; Shen, Y.; Zhang, Z. Balanced stress fracturing theory and its application in platform well fracturing during unconventional oil and gas development. *Energy Rep.* **2022**, *8*, 10705–10727. [CrossRef]
8. Lu, C.H.; Jiang, H.Q.; Yang, J.L. Simulation and optimization of hydraulic fracturing in shale reservoirs: A case study in the Permian Lucaogou formation, China. *Energy Rep.* **2022**, *8*, 2558–2573. [CrossRef]
9. Toma, S.H.; Santos, J.J.; Silva, D.G.; Huila, M.F.G.; Toma, H.E.; Araki, K. Improving stability of iron oxide nanofluids for enhanced oil recovery: Exploiting wettability modifications in carbonaceous rocks. *J. Pet. Sci. Eng.* **2022**, *212*, 110311. [CrossRef]
10. Yang, L.; Jin, Z.J. Global shale oil development and prospects. *China Pet. Explor.* **2019**, *24*, 553–559. (In Chinese)
11. Jin, Z.J.; Zhu, R.K.; Liang, X.P.; Shen, Y.Q. Several issues worthy of attention in current lacustrine shale oil exploration and development. *Pet. Explor. Dev.* **2021**, *48*, 1471–1484. [CrossRef]
12. Bai, G.; Qiu, H.H.; Deng, Z.Z.; Wang, W.Y.; Chen, J. Distribution and main controls for shale oil resources in USA. *Pet. Geol. Exp.* **2020**, *42*, 524–532. (In Chinese)
13. Jin, Z.J.; Gao, D.L.; Jia, C.Z. *Research on China's Shale Oil Development Strategy*; Faculty of Chinese Academy of Sciences: Beijing, China, 2019. (In Chinese)
14. Boswell, R.; Yoneda, J.; Waite, W.F. India national gas hydrate program expedition 02 summary of scientific results: Evaluation of natural gas-hydrate-bearing pressure cores. *Mar. Pet. Geol.* **2019**, *108*, 143–153. [CrossRef]
15. Chen, B.B.; Liu, Z.Y.; Sun, H.R.; Zhao, G.J.; Sun, X.; Yang, M.J. The synthetic effect of traditional-thermodynamic-factors (temperature, salinity, pressure) and fluid flow on natural gas hydrate recovery behaviors. *Energy* **2021**, *233*, 121147. [CrossRef]
16. Ruppel, C.D.; Kessler, J.D. The interaction of climate change and methane hydrates. *Rev. Geophys.* **2017**, *55*, 126–168. [CrossRef]
17. McConnell, D.R.; Zhang, Z.; Boswell, R. Review of progress in evaluating gas hydrate drilling hazards. *Mar. Pet. Geol.* **2012**, *34*, 209–223. [CrossRef]
18. Sun, X.; Luo, T.T.; Wang, L.; Wang, H.J.; Song, Y.C.; Li, Y.H. Numerical simulation of gas recovery from a low-permeability hydrate reservoir by depressurization. *Appl. Energy* **2019**, *250*, 7–18. [CrossRef]
19. Ye, H.Y.; Wu, X.Z.; Li, D.Y.; Jiang, Y.J.; Gong, B. A novel thermal stimulation approach for natural gas hydrate exploitation—The application of the self-entry energy compensation device in the Shenhu sea. *J. Nat. Gas Sci. Eng.* **2022**, *105*, 104723. [CrossRef]
20. Aghajari, H.; Moghaddam, M.H.; Zallaghi, M. Study of effective parameters for enhancement of methane gas production from natural gas hydrate reservoirs. *Green Energy Environ.* **2019**, *4*, 453–469. [CrossRef]
21. Feng, G.; Kang, Y.; Sun, Z.D.; Wang, X.C.; Hu, Y.Q. Effects of Supercritical CO_2 adsorption on the mechanical characteristics and failure mechanisms of shale. *Energy* **2019**, *173*, 870–882. [CrossRef]
22. Jadhawar, P.; Yang, J.H.; Chapoy, A.; Tohidi, B. Subsurface carbon dioxide sequestration and storage in methane hydrate reservoirs combined with clean methane energy recovery. *Energy Fuel* **2021**, *35*, 1567–1579. [CrossRef]
23. Fan, S.S.; Yu, W.Y.; Yu, C.; Wang, Y.H.; Lang, X.M.; Wang, S.L.; Li, G.; Huang, H. Investigation of enhanced exploitation of natural gas hydrate and CO_2 sequestration combined gradual heat stimulation with CO_2 replacement in sediments. *J. Nat. Gas Sci. Eng.* **2022**, *104*, 104686. [CrossRef]
24. Viney, N.; Post, D.A.; Crosbie, R.S.; Peeters, L.K.M. Modelling the impacts of future coal mining and coal seam gas extraction on river flows: A methodological framework. *J. Hydrol.* **2021**, *596*, 126144. [CrossRef]
25. *BP Energy Outlook—2017 Edition*; BP p.l.c.: Houston, TX, USA, 2017.
26. Wang, G.F.; Xu, Y.X.; Ren, H.W. Intelligent and ecological coal mining as well as clean utilization technology in China: Review and prospects. *Int. J. Min. Sci. Technol.* **2019**, *29*, 161–169. [CrossRef]

27. Lv, H.Y.; Cheng, Z.B.; Liu, F. Study on the mechanism of a new fully mechanical mining method for extremely thick coal seam. *Int. J. Rock Mech. Min.* **2021**, *142*, 104788. [CrossRef]
28. Peng, S.S.; Du, F.; Cheng, J.Y.; Li, Y. Automation in U.S. longwall coal mining: A state-of-the-art review. *Int. J. Min. Sci. Technol.* **2019**, *29*, 151–159. [CrossRef]
29. Liu, H.F.; Wang, Y.J.; Pang, S.; Wang, X.F.; He, J.G.; Zhang, J.X.; Alfonso, R.D. Mining footprint of the underground longwall caving extraction method: A case study of a typical industrial coal area in China. *J. Hazard. Mater.* **2022**, *425*, 127762. [CrossRef] [PubMed]
30. Poormirzaee, R.; Hosseini, S.; Tagh, R. Smart mining policy: Integrating fuzzy-VIKOR technique and the Z-number concept to implement industry 4.0 strategies in mining engineering. *Resour. Policy* **2022**, *77*, 102768. [CrossRef]
31. Li, F.; Li, L.X. Key technologies of big data processing platform construction for coal mining. *J. China Coal Soc.* **2019**, *44* (Suppl. 1), 362–369. (In Chinese)
32. Yuan, L. Scientific conception of precision coal mining. *J. China Coal Soc.* **2017**, *42*, 1–7. (In Chinese)

Article

A Review of Intelligent Unmanned Mining Current Situation and Development Trend

Kexue Zhang [1,2,3,4,*,†], Lei Kang [1,2,*,†], Xuexi Chen [1,2], Manchao He [3], Chun Zhu [5] and Dong Li [1,2]

[1] Institute of Intelligent Unmanned Mining, North China Institute of Science and Technology, Beijing 101601, China; xuexichen@ncist.edu.cn (X.C.); 2019013771ld@ncist.edu.cn (D.L.)
[2] Hebei Provincial Key Laboratory of Mine Intelligent Unmanned Mining Technology, North China Institute of Science and Technology, Beijing 101601, China
[3] State Key Laboratory for Geomechanics & Deep Underground Engineering, China University of Mining and Technology (Beijing), Beijing 100083, China; hemanchaocumtb@163.com
[4] China Coal Research Institute, Beijing 100013, China
[5] School of Earth Sciences and Engineering, Hohai University, Nanjing 210098, China; zhu.chun@hhu.edu.cn
* Correspondence: zhangkexue@ncist.edu.cn (K.Z.); 201908522470kl@ncist.edu.cn (L.K.); Tel.: +86-133-6603-0731 (K.Z.)
† These authors contributed equally to this work.

Abstract: Intelligent unmanned mining is a key process in coal mine production, which has direct impact on the production safety, coal output, economic benefits and social benefits of coal mine enterprises. With the rapid development and popularization of 5G+ intelligent mines and coal mine intelligent equipment in China, the intelligentization of intelligent unmanned mining has become an important research topic. Especially with the promulgation of some Chinese policies and regulations, intelligent unmanned mining technology has become one of the key technologies of coal mine production. To understand the connotation, status quo and development trends of intelligent unmanned mining, this paper takes the relationship between key technologies and engineering application of intelligent unmanned mining in China as the perspective. It is proposed that the intelligent unmanned mining technology is in the whole process of working face mining. A research structure of unmanned follow-up operation and safe patrol is changing to the mode of intelligent adaptive mining, followed by the basic concepts and characteristics of intelligent unmanned mining. Relevant researches that maybe beneficial to the proposed research content are reviewed in four layers, which include basic theory, key technology, mining mode, and overall design system theory and technology. Finally, the current intelligent unmanned mining mode and future trends in this field in China are summarized.

Keywords: intelligent mine; smart mining; intelligent working face; intelligent control; adaptive; robotic technology

Citation: Zhang, K.; Kang, L.; Chen, X.; He, M.; Zhu, C.; Li, D. A Review of Intelligent Unmanned Mining Current Situation and Development Trend. *Energies* **2022**, *15*, 513. https://doi.org/10.3390/en15020513

Academic Editor: Isabel Jesus

Received: 7 December 2021
Accepted: 7 January 2022
Published: 12 January 2022

Publisher's Note: MDPI stays neutral with regard to jurisdictional claims in published maps and institutional affiliations.

1. Introduction

The application of artificial intelligence, industrial Internet, cloud platform, big data, robot, 5G and other advanced technologies in the field of intelligent unmanned coal mining has promoted the innovation and development of intelligent unmanned coal mining in China's coal industry [1–3].

In order to implement *The guiding opinions on accelerating the intelligent development of coal mines*, jointly issued by the National Development and Reform Commission, the Energy Bureau, the Ministry of Emergency, the Coal Supervision Bureau, the Ministry of Industry and Information Technology, the Ministry of Finance, the Ministry of Science and Technology, and the Ministry of Education [4], as of November 2020, five provinces in China have given specific implementation plans or opinions on the intelligent construction of coal mines. It includes *The Implementation Plan of Intelligent Coal Mine Construction in*

Henan Province, The Implementation Plan of Intelligent Coal Mine Construction in Shandong Province, The Implementation Opinions of Intelligent Coal Mine Construction in Shanxi Province, The Implementation Plan of Intelligent Coal Mine Development in Guizhou Province (2021–2025) and *The Implementation Opinions of Accelerating Intelligent Coal Mine Development in Yunnan Province.* On 8 December 2020, Shanxi Provincial Energy Bureau issued the Evaluation Method for Intelligent Construction of Coal Mines in the Province (Trial) and the Basic Requirements and Scoring Method for Intelligent Construction of Coal Mines in the Province (Trial). The promulgation and implementation of the implementation plans or opinions for the intelligent construction of coal mines will provide policy basis for the development of intelligent unmanned mining [5–8].

Intelligent unmanned mining is a key process in the production of coal mines, metal mines, non-metal mines and other mining industries, which directly affects the safety, output and benefit [9,10]. However, although intelligent unmanned mining breaks the traditional idea of controlling the target with single machine position on the basis of manual operation, it improves the degree of automation of fully-mechanized mining, liberates workers from the working face, and realizes the reduction of personnel in working face mining. However, there are also limitations: the level of intelligent construction of coal mines is unbalanced; the level of regional intelligentization with good geological conditions in western China is higher; and the level of regional intelligentization with poor geological conditions in southwestern China is lower [11].

With the rapid development of intelligent unmanned mining technology, a large amount of real-time data and historical data will be generated during the coal mining process, how to collect and monitor those data and experience knowledge, how to realize a higher level of intelligent mining through key technologies, and how to realize unmanned mining of working face. These problems have gradually become research hotspots [12].

Therefore, this paper constructs the research system architecture of the intelligent unmanned mining in the technology-driven mode, introduces the basic concepts and characteristics of the intelligent unmanned mining, and explores the key technologies of intelligent unmanned mining, and summarizes the intelligent unmanned mining mode. Combined with the application of the intelligent unmanned mining in typical coal mines, it points out the problems existing in the research of intelligent unmanned mining in coal mines in China. Finally, the development trend of China's intelligent unmanned mining is prospected.

2. The Development History of Unmanned Mining Technology

Intelligent unmanned mining is developed on the basis of mechanization, automation, systematization and visualization. At present, unmanned mining technology mainly includes auger unmanned mining, coal planer unmanned mining, comprehensive mechanized unmanned mining, and intelligent unmanned fully-mechanized mining. Unmanned coal mining with auger rigs is generally suitable for the mining of thin coal seams or very thin coal seams, and some coal mines are also used for the mining of side coal in the open pit [13]. At present, three-axis auger has been widely used in coal mining at home and abroad. Unoperated coal plough mining is generally applicable to thin coal seam with coal seam inclination less than 25° and a stable slope of working face, extremely thin coal seam, or coal seam with high gas content [14]. Compared with auger drills, coal ploughs are more stable and reliable, low cost and with a high degree of automation. Intelligent unmanned mining is generally used for the mining of medium and thick coal seams. Through the intelligent control system of the working face, the use of visual remote monitoring is used to realize the intelligent operation of coal mining, support, and coal transportation at the working face. Intelligent fully-mechanized caving for extra-thick coal seams is mainly achieved through the use of ultra-large mining height top coal caving hydraulic supports, intelligent coal caving systems, intelligent control of key technologies, etc., to achieve intelligent mining [15,16]. Comprehensive mechanized unmanned coal mining is of great significance to the development of intelligent unmanned mining. With the development of integrated mechanized unmanned coal mining, new technologies have emerged, including intelligent unmanned mining basic platforms, intelligent hydraulic support systems [17], intelligent mining models, and underground autonomous vehicles [18], etc., which have greatly promoted the development of intelligent unmanned mining developing. The fully-mechanized coal mining face is transforming from mechanization to automation, eventually on the basis of traditional fully-mechanized mining technology, the use of hydraulic supports, coal winning machines, scraper conveyors, and other mining equipment with perception, decision-making and execution capabilities are used to automate the control system in the core, and visual remote monitoring is used as the means to realize the safe and efficient mining mode of "unmanned follow-up operation and manned safety patrol" in the whole process of coal mining at the working face [19].

3. Present Situation of Intelligent Unmanned Mining in China

China's intelligent unmanned mining has developed rapidly. Up to now, more than 300 intelligent coal mining working faces have been built. Coal production has realized a historic leap from manual operation to mechanization, automation, informationization and intelligence. With the application of new technologies and equipment such as 5G+ intelligent mines, coal mine robots, and coal mine intelligent equipment in intelligent unmanned mining, the intelligent development of coal mine in China has ushered in new development opportunities [20].

Chinese scholars have done a lot of research work in the field of intelligent unmanned mining, and have achieved remarkable results. The details are as follows: Wang Guofa et al. [21] proposed the construction of eight intelligent systems based on the big data application center, pointed out the basic structure and principles of the design and construction of coal mine intelligent systems, and proposed the research direction of intelligent unmanned mining including precise geological information. This included system and mining and mining detection technology and equipment, smart coal mine Internet of Things technology and equipment, roadway intelligent rapid excavation technology and equipment, intelligent unmanned mining key technology and equipment, coal mine robot technology and product research and development. Li Shoubin [22] summarized the four stages of intelligent mining, analyzed the production characteristics and technical requirements of different stages, and proposed a control theory model under the mode of intelligent self-adaptive mining technology, making use of key new technologies to advance toward intelligent mining. Fan Jingdao [23] analyzed the technical problems existing in intelligent fully-mechanized mining of the working face with large mining height, and realized the smooth operation of intelligent mining with large mining height by using high-efficiency mining process, anti-slake control technology, bottom soft intelligent control technology and roof broken intelligent control technology. Zhang Kexue [24,25] studied the current situation and development trend of intelligent mining technology in a fully-mechanized working face, divided the intelligent unmanned mining technology of comprehensive excavating working face into intelligent unmanned mining with visual and remote intervention and intelligent unmanned mining with self-adaptation, and proposed the key technology of intelligent unmanned mining of comprehensive excavating of the working face. The intelligent mining procedure, control system and technical route of comprehensive mining face are obtained, and the method of applicability evaluation of intelligent unmanned mining face is put forward. Based on the development status of visual remote intervention, robot-assisted patrol inspection and inertial navigation technology, Huang Zenghua et al. [26] proposed the basic architecture of intelligent coal mining system with four dimensions of "perception, decision, execution, operation and maintenance", and identified the technical direction of key technologies of intelligent coal mining to be breakthrough. Wang Cunfei et al. [27] proposed the global model of intelligent unmanned mining transparent working face, including the perception layer, the transmission layer, the information layer and the intelligent application layer. Key technologies for the construction of transparent working face were analyzed, and the technical bottlenecks existing in the construction of transparent working face were presented. Wang Jinhua et al. [28] put forward the next work focus of intelligent technology and equipment in fully-mechanized mining face by studying the intelligent production mode and the status quo of intelligent technology and equipment in fully-mechanized mining face.

4. Basic Theory

4.1. Definition of Intelligent Unmanned Mining

Intelligent unmanned mining refers to the application of advanced technologies such as 4G or 5G communications, Internet of Things, cloud computing, big data, artificial intelligence, and other advanced technology, adopted with independent awareness, independent decision-making and control execution ability of the coal winning machine, hydraulic support, scraper conveyor, reprint machine, crusher, belt conveyor, mining, and transportation equipment, such as an intelligent integrated intelligent control system at the core. By means of visual remote monitoring, a safe and efficient comprehensive mechanized coal mining method with intelligent operation process (self-adaptation of working condition and collaborative control of working procedure) or one-button start-stop operation mode (including unmanned follow-up operation and manned safety patrol) can be realized in coal mining, support and transport of the working face.

4.2. Multi-Source Heterogeneous Data Model

With the rapid development of intelligent unmanned technology and intelligent equipment, the amount of monitoring data in coal mines has increased sharply, and how to collect and process data accurately, quickly, and conveniently has become particularly important

The multi-source heterogeneous data model [29,30] is a bottom-up method; by extracting different types, different attributes, different levels, and different statuses of the monitoring data, according to the actual information needs, the unified description of the multi-source heterogeneous data, and based on the one-to-one mapping relationship between various monitoring data, a mine information sharing service platform is established to provide key data for coal production. At present, including coal mine big data analysis and decision-making technology, fault self-diagnosis technologies conduct the collection and processing of monitoring data.

Based on the multi-source heterogeneous data model, big data fusion technology, model digitization technology, analysis and decision technology, and machine learning technology, Zhang Kexue established a set of transparent working face intelligent mining systems that can be "forecasted, predicted, and controlled" in a big data analysis and decision-making platform [31]. At present, the multi-source heterogeneous data model has been successfully used in the intelligent analysis and decision-making of the coal mining face mining process. At the same time, it also has broad application prospects in other aspects involving monitoring data, such as intelligent and precise mining of coal mines, intelligent video monitoring of coal mines, and coal mine robots.

4.3. Transparent Geological Prediction Model

Based on the data of fine roadway measurements, borehole detection [32–34] and channel wave seismic exploration, using implicit iterative interpolation algorithm and data volume error analysis and other geological modeling techniques to carry out three-dimensional geological modeling, discretize the cutting path of the shearer, and establish the digital coal seam model builds a transparent geological prediction model through continuous updating of realistic data. The key technologies of the transparent geological prediction model include comprehensive detection technology, geological modeling technology, "CT" slice technology, and dynamic update technology [35].

(1) Comprehensive detection technology

The fine measurement of the roadway revealed the geological structure, coal seam floor and coal seam undulations. Borehole detection obtains the spatial position information of the roof and floor of the coal seam in the middle of the working face. Channel wave seismic exploration revealed the hidden structure and coal thickness distribution of coal seams.

The formula for calculating the spatial attitude parameters of borehole detection is as follows:

$$tan\theta = \frac{-Gy}{\sqrt{Gx^2 + Gz^2}} \cdot \quad (1)$$

$$tan\psi = \frac{-Gy}{Gz} \quad (2)$$

$$tan\sigma = \frac{-G0(HzGx - HxGz)}{Hy(Gx^2 + Gz^2) - Gy(HzGz + HxGx)} \quad (3)$$

In the Equations (1)–(3), Gx, Gy, Gz represent the three components of the gravity acceleration sensor. Hx, Hy, Hz represent the three components of the magnetoresistive sensor, Hx, Hy, Hz indicate that the three-axis magnetoresistive sensor measures the different spatial magnetic field components of the geomagnetic field during borehole detection, and outputs them after amplification, filtering and A/D conversion. $G0$ is the local gravity acceleration value, θ is the inclination angle, ψ is the tool surface, σ is the azimuth angle, and the inverse trigonometric function can be solved to obtain the borehole detection space attitude parameters.

(2) Geological modeling technology

Three-dimensional geological modeling includes two parts: structural modeling and attribute modeling. For the current development stage (used for smart working face), the construction model can meet the demand. The key technologies of geological modeling technology include implicit iterative interpolation algorithm and data volume error analysis.

The interpolation principle of the implicit iterative interpolation algorithm is to establish a network of interconnections through a discrete natural body model. If the value of the point on the network meets certain constraints, the value of the unknown node can be solved by solving the linear equations obtained. Aiming at the problem of estimating the (φ) value on the grid node, the implicit iterative interpolation algorithm establishes the objective function of calculating the optimal solution of the grid node:

$$R^*(\varphi) = R(\varphi) + \rho(\varphi) \quad (4)$$

In Formula (4), $R(\varphi)$ is the global roughness function, and $\rho(\varphi)$ is the linear constraint violation function.

By minimizing the objective function $R^*(\varphi)$ two goals are achieved: 1. Minimizing the global roughness function $R(\varphi)$, so that the function value on any node is as close to the mean value of the node in the field as possible, that is, the value of each node the value (φ) is as smooth as possible; 2. Convert the original sampled data into linear constraints defined on some nodes to minimize the violation degree $\rho(\varphi)$ of the linear constraints, that is to say, maximize the compliance with the linear constraints, so as to make the value (φ) of the related nodes approximate the sampling data as much as possible.

The theoretical formula of the implicit iterative modeling algorithm:

$$R_{MSE} = \sqrt{\frac{1}{n}\sum_{i=1}^{n}(x_i - x_0)^2} \quad (5)$$

$$M_{AE} = \frac{1}{n}\sum_{i=1}^{n}\sqrt{(x_i - x_0)^2} \quad (6)$$

In Formula (5), R_{MSE} represents the root mean square error, which reflects the estimation sensitivity and extreme value effect of using sample data; in Formula (6), M_{AE} represents the average absolute error, which represents the possible error range of the estimated predicted value; x_0 is the measured value of the i-th point, $\overline{x_0}$ is the average of the measured values, x_i is the predicted value of the i-th point, $\overline{x_i}$ is the average of predicted values, and n is the number of inspection points.

(3) "CT" slice technology

Due to the insufficient geological adaptability and stability of the current intelligent mining system, a "CT" slicing technology is proposed, which is to cut the digital model of the coal seam to be mined according to the cutting plan, and then according to the cutting surface and the cutting surface of the digital coal seam model. Intelligent mining requires optimizing the cutting path and parameters of the shearer, and controlling the shearer to mine according to the planned cutting path.

The principle of the discretization of the cutting path is to assume that the projection curve is scattered into n line segments. For the i-th line segment, the coordinates of the two ends are (x_i, y_i) and (x_{i+1}, y_{i+1}), respectively, and the straight line equation between the two points is calculated as:

$$y = k_i x + b_i \tag{7}$$

In Formula (7), $x \in [\ \min(x_i, x_{i+1}), \quad \max(x_i, x_{i+1})\ ]$, $k_i = \frac{y_i - y_{i+1}}{x_i - x_{i+1}}$, $b_i = y_i - x_i \frac{y_i - y_{i+1}}{x_i - x_{i+1}}$, and $i = 1, 2, \cdots, n$.

Calculate the intersection of the line equation $y = k_i x + b_i$ in the interval B and the grid line. Perform the above two steps on all the line segments to obtain the approximate projection point sequence of the planned cutting route on the two-dimensional plane.

The cutting principle of the coal seam digital model is as follows: For the j-th projection point (x_j, y_j), set the neighborhood parameter r, and search for all the points where the top grid and bottom grid point plane coordinates fall within the neighborhood area $\{x_j + r < x < x_j + r, y_j + r < y < y_j + r\}$.

The elevations z_{1i} and z_{2i} of the top and bottom plates corresponding to the j-th projection point (x_j, y_j) can be determined according to two methods: the closest distance method and the distance weighted method. The closest distance method is to take the elevation value of the coal roof and floor corresponding to the grid point with the closest distance to the two-dimensional plane of the projection point in the neighborhood. The distance weighting method calculates the plane distance from each grid point in the neighborhood to the projection point (x_j, y_j), assigns the top and bottom plate elevation weights according to the reciprocal of the plane distance, and calculates the top and bottom plate elevations z_{1i} and z_{2i} corresponding to the projection point (x_j, y_j) by weighting, calculated as follows:

$$z_{1i} = \sum_{m=1}^{M} W_m Z_{tm} \tag{8}$$

$$z_{2i} = \sum_{m=1}^{M} W_m Z_{bm} \tag{9}$$

In Formulas (8) and (9), Z_{tm}, Z_{bm}, and W_m are the elevation of the top plate, the elevation of the bottom plate corresponding to the m-th grid point in the neighborhood, and the weighted weight at that point. The calculation formula for W_m is:

$$\begin{cases} k(\frac{1}{l_1} + \frac{1}{l_2} + \cdots + \frac{1}{l_m} + \cdots + \frac{1}{l_M}) = 1 \\ W_m = k\frac{1}{l_m} \end{cases}, 1 \leq m \leq M \tag{10}$$

Figure 1 is a simple schematic diagram of Formulas (7)–(9). Roof represents the roof of the coal seam, Bottom plate represents the floor of the coal seam, x_j represents the coordinates in the x-axis direction of the j-th projection point, and y_j represents the coordinates in the y-axis direction of the j-th projection point. Zt_m, Z_{bm} are the elevation of the top plate and the elevation of the bottom plate corresponding to the m-th grid point in the neighborhood.

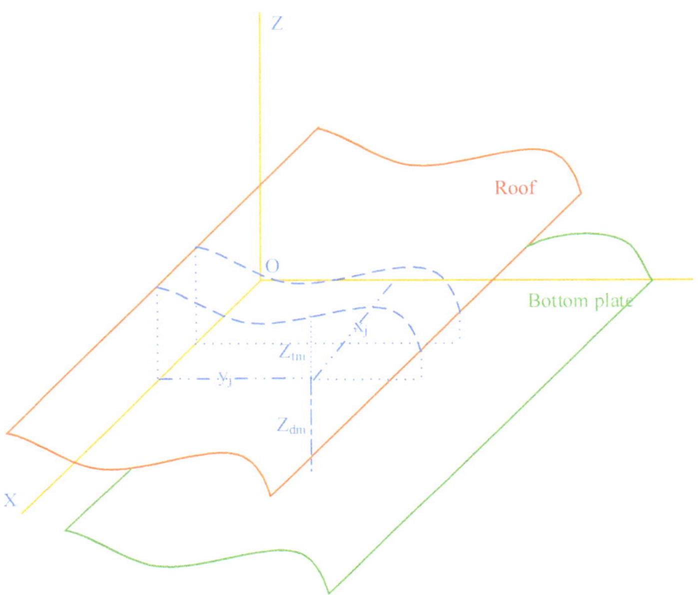

Figure 1. Schematic diagram of "CT" slicing technology.

(4) Dynamic update technology

The data update of the model is that the coal seam exposed during the mining process is re-realistic and combined with the newly generated geological data during the production process to update the model [36–38]. With the continuous increase of data, the accuracy of the model is also continuously improved.

Geological prediction models play an important role in the field of intelligent unmanned mining [39]. Take Huangling Mining as an example. On 15 September 2020, the "Intelligent Mining Technology Based on Dynamic Geological Model Big Data Fusion Iterative Planning Control Strategy" project of Huangling Mining Company passed the scientific and technological achievements appraisal of the China Coal Industry Association. The project innovatively developed the 3D modeling technology of an implicit iterative algorithm, built a high-precision geological model of fully-mechanized mining face, and proposed a comprehensive detection and analysis method based on fully-mechanized mining face as the research object to accurately predict abnormal structures in the working face, providing geological guarantee for the intelligent mining. Based on the data of geological realism, borehole radar detection and channel wave seismic exploration, a static geological model of a fully-mechanized mining face is constructed. They developed implicit iterative modeling and dynamic update algorithms to achieve dynamic updates of the static geological model. The accuracy error of the updated model was within 150 mm, and real-time correction of cutting parameters was realized, forming a 22 planned cutting process for the shearer and 14 planning and control processes of hydraulic support [40].

5. Key Technologies for Intelligent Unmanned Mining

The key technologies of intelligent unmanned mining are divided into intelligent automation technology, intelligent control technology, intelligent monitoring technology, intelligent precise positioning technology and other intelligent technologies. Intelligent automation technology includes shearer memory cutting technology and intelligent automatic rapid tunneling technology. Intelligent control technology includes centralized control technology for fully-mechanized mining equipment, intelligent integrated liquid supply control technology, coal flow load feedback coal mining control technology [41] and remote-control technology. Intelligent surveillance technology includes intelligent video surveillance technology and intelligent video positioning and tracking technology. Intelligent precise positioning technology includes personnel precise positioning technology and equipment precise positioning technology. Other intelligent technologies include intelligent auxiliary transportation technology, intelligent ventilation technology, intelligent sorting technology and intelligent underground robot technology.

5.1. Shearer's Memory Cutting Technology

Figure 2 shows the schematic diagram of the shearer memory cutting technology. When the shearer is located at A-A in the figure, it means that the shearer drum is normally cutting coal. When the shearer is located at B-B in the figure, it means that the position of the shearer drum is lower. Through the shearer memory cutting technology, the height of the drum is automatically adjusted to the normal coal cutting state C-C of the shearer drum. When the shearer is located at D-D in the figure, it means that the position of the shearer drum is on the upper side. Through the shearer memory cutting technology, the height of the drum is automatically adjusted to the normal coal cutting state E-E of the shearer drum. Through a variety of sensors in the body of the shearer to realize the shearer's mining height, speed and other data acquisition, the shearer self-positioning device was developed to achieve the shearer cutting process of automatic control. Additionally, the memory is carried out in the control program database to realize the learning of "demonstration knife", and realize the shearer automatically cutting coal according to the memorized curve and technology in the next cycle process, and finally realize the memory cutting [42]. Fan Qigao et al. [43] corrected the grey model by using the Markov chain state probability matrix, and obtained the adaptive adjustment value of shearer cutting height, which greatly improved the automation level of the shearer. Zhang Lili et al. [44,45] used a genetic algorithm or particle swarm algorithm to optimize the memory cutting path of shearer. The experimental results show that genetic algorithm or particle swarm algorithm can quickly and effectively realize the path optimization of the shearer under a complex geological environment, which is beneficial to the memory cutting and automatic control of the shearer.

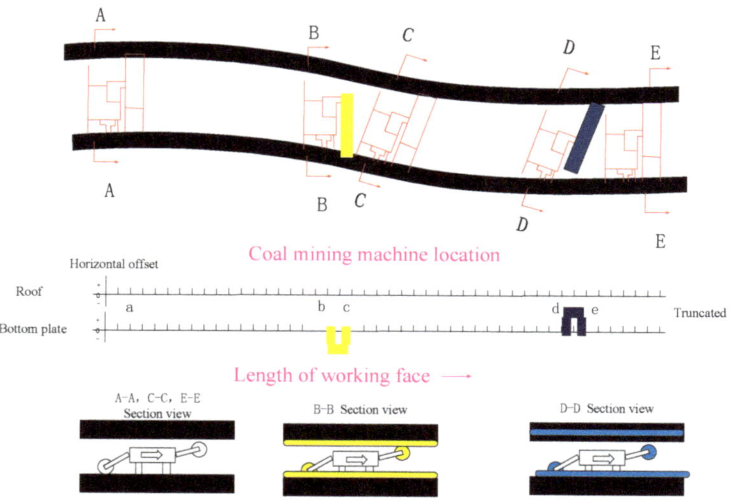

Figure 2. Schematic diagram of shearer memory cutting technology: A-A, C-C and E-E means that the shearer drum is normally cutting coal, B-B means that the position of the shearer drum is lower, D-D means that the position of the shearer drum is on the upper side.

5.2. Intelligent Control Technology of Hydraulic Support

Figure 3 shows the structure diagram of the Inertial Navigation Straight Finding System. Figure 4 shows the schematic diagram of the automation of hydraulic support and machine. The left side of Figure 4 shows the shearer that is cutting coal. At this time, the red line indicates that the scraper conveyor is in a curved state. The automatic follow-up of the hydraulic support is used to automatically straighten it, so that the straightening of the scraper conveyor on the working face is as shown in Figure 4. With the green line on the right, the working face has moved forward. At present, the intelligent hydraulic support with independent intellectual property rights in China has been equipped with the action of following the automatic advancement and automatic support of the shearer, and has realized some intelligent functions, including automatic frame shifting, automatic follow-up, automatic incline adjustment, automatic straightening, intelligent liquid supply, self-testing mine pressure, automatic pressure compensation and centralized control of roadways, etc. [46]. The intelligent control technology of the hydraulic support is real-time control through remote intervention of the hydraulic support, using the monitoring data of the hydraulic support and the visual monitoring technology of the video follower to realize the automatic transfer of the hydraulic support, automatic follow-up, automatic incline adjustment, straightening and other functions.

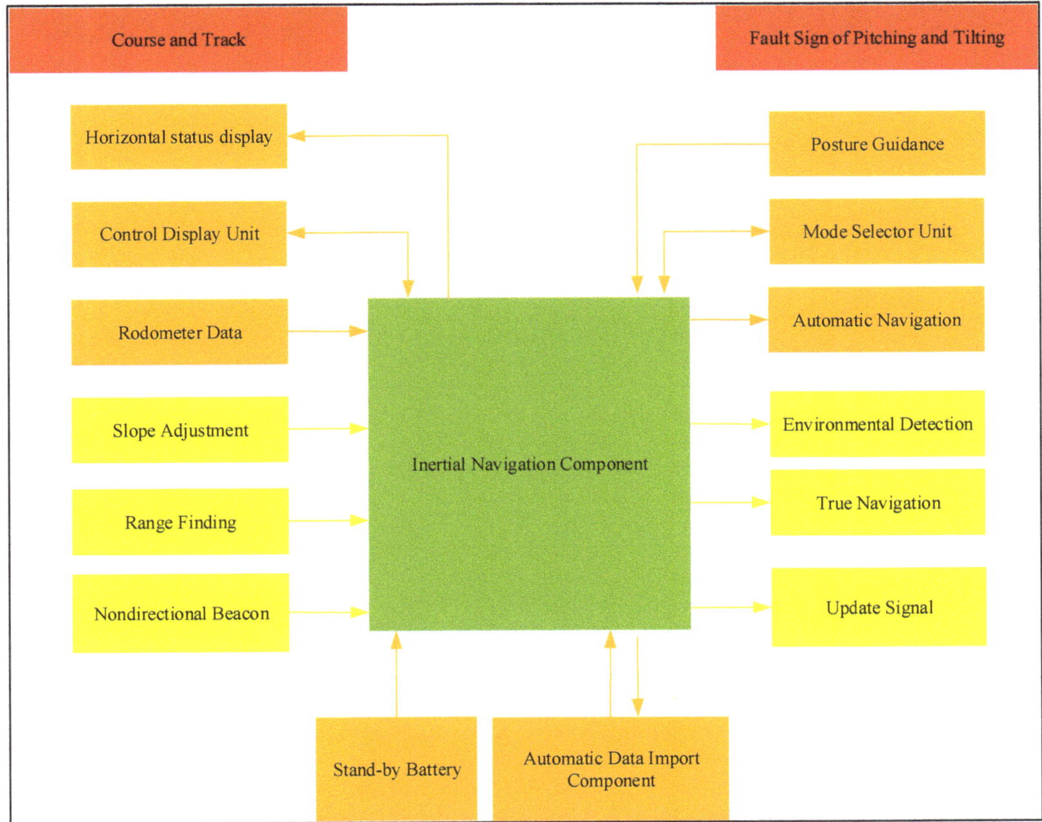

Figure 3. Structure diagram of the Inertial Navigation Straight Finding System.

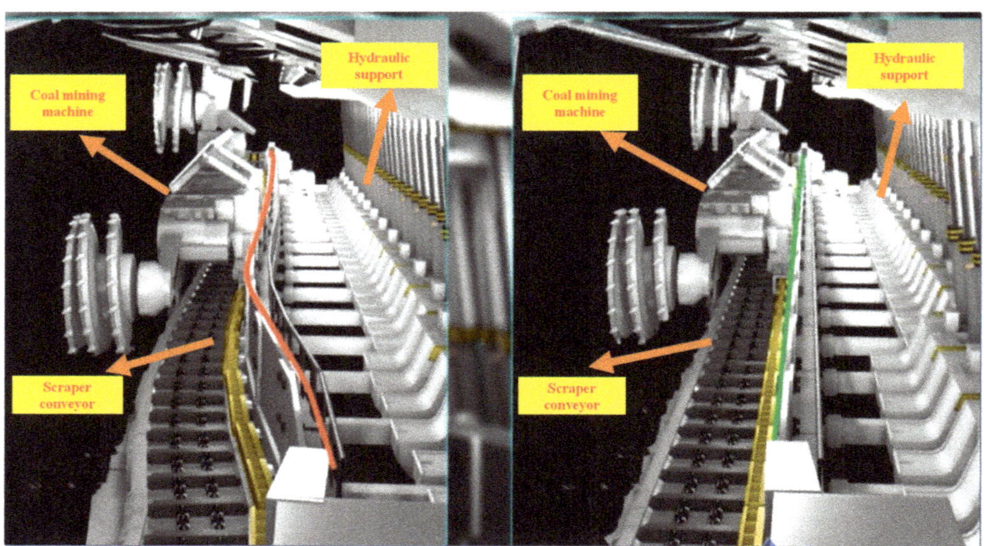

Figure 4. Schematic diagram of hydraulic support and machine automation.

5.3. Centralized Control Technology of Fully-Mechanized Mining Equipment

Figure 5 shows the process control model of fully mechanized mining equipment. Figure 6 shows the picture of the centralized control software for fully-mechanized mining equipment. Figure 6a is the coal wall monitoring system, which collects coal wall conditions in real time through video; Figure 6b is the hydraulic pump monitoring system, which realizes the visualization of hydraulic data of the hydraulic support, and Figure 6c is the coal shearer monitoring system, which can monitor the shearer drum in real time. In the roof condition, Figure 6d shows that the shearer detection system monitors the relevant data of the shearer, such as shearer speed, traction current, load cutting temperature, etc., to clarify the normal value interval and early warning and alarm thresholds of various monitoring data. Various failures of the shearer equipment, including electrical control, mechanical transmission, hydraulic transmission, etc., are corresponding solutions to the failure; Figure 6e shows the statistics of the use of the shearer. The use of the shearer is recorded and the maintenance is designed. The arrangement of the plan is based on the relevant rules and requirements of the maintenance work. Figure 6f is a 3D scene model of the stope, which supports the location mark of the shearer equipment in the 3D scene and the fine modeling of the equipment, and realizes the three-dimensional visualization of the coal mining face. Through a set of centralized control system for fully-mechanized mining equipment, the centralized automation of "three machines" (scraper conveyor, shearer, hydraulic support), transfer machine, crusher, pump station and other fully-mechanized mining equipment at fully-mechanized mining face is realized control. The centralized control technology of fully-mechanized mining equipment also involves multiple coal mine systems such as the power supply system, liquid supply system, monitoring system and communication system. Huang et al. [47] proposed and designed a set of centralized control systems for fully-mechanized working face equipment, which can integrate control, Ethernet, video, wireless, communication, monitoring and other technologies. It realized the centralized automatic control of fully-mechanized equipment in the roadway monitoring center, and was successfully applied in the 4109 working face of the No. 1 Mine of the Pingshuo Group. Based on the analysis of the necessity of developing electro-hydraulic control technology, Song et al. [48] reviewed the functional requirements and technical characteristics of electro-hydraulic control technology at different stages of development. Then, the current research status of electro-hydraulic control technology at home and abroad is introduced, and the development direction of electro-hydraulic control technology of hydraulic support in fully-mechanized automation working face is prospected. Taking Gaozhuang Coal Industry Co., Ltd. as the research object, Liu et al. [49] designed the integrated control system scheme of fully-mechanized mining equipment, analyzed the automation level of fully-mechanized mining equipment in working face, and studied the control function, data acquisition function, data storage function and alarm function of fully-mechanized mining equipment, which improved the labor efficiency of enterprises, ensured safe production, and increased economic benefits.

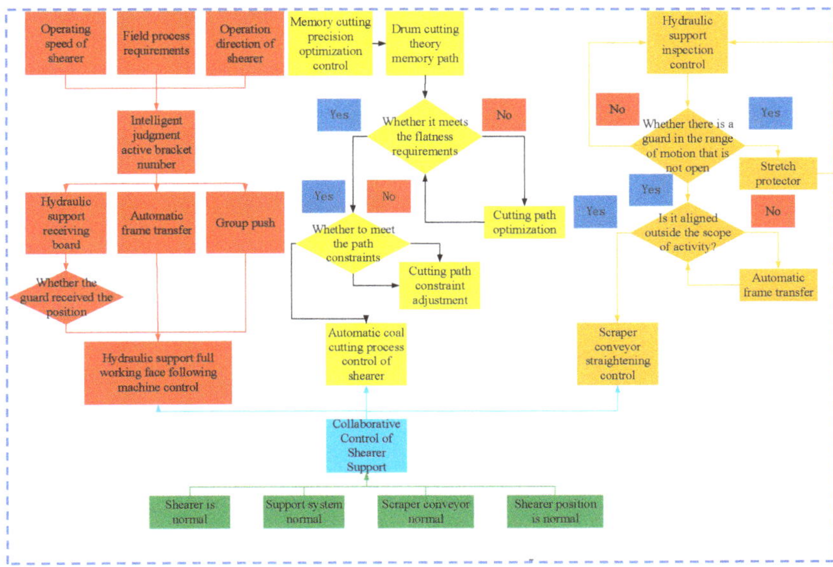

Figure 5. Process control model of fully mechanized mining equipment.

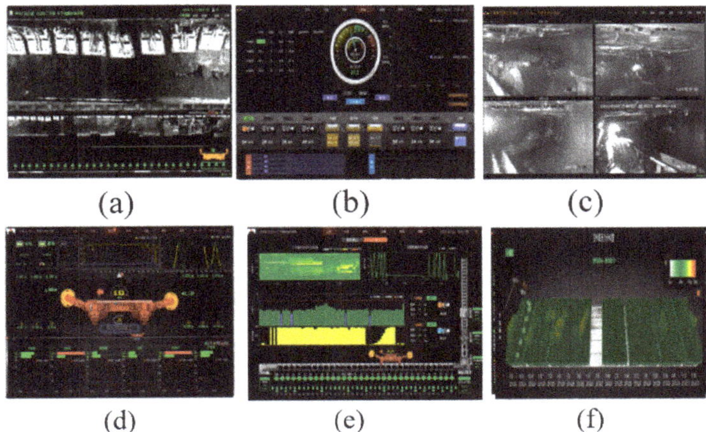

Figure 6. Centralized control software screen of comprehensive mining equipment: (**a**) coal wall monitoring system, (**b**) hydraulic pump monitoring system, (**c**) coal shearer monitoring system, (**d**) hearer detection system monitors the relevant data of the shearer, (**e**) statistics of the use of the shearer, (**f**) 3D cene model of the stope.

5.4. Intelligent Video Surveillance Technology

Figure 7 shows the schematic diagram of intelligent video surveillance technology. Intelligent video surveillance technology is the key technology to realize intelligent unmanned mining. In addition to the resolution of high-definition and above, auto-focusing function, and meeting the requirements of mine-used intrinsic safety design, the intelligent video surveillance system should have the functions of target detection, target recognition, and behavior analysis. Chen Guiping [50] proposed the use of intelligent visual analysis and pattern recognition combined with intelligent analysis and early warning technology to achieve intelligent operation of coal mine video surveillance, through the analysis of abnormal monitoring screen to achieve active early warning, and effectively overcome the shortcomings of traditional video surveillance system. Cheng et al. [51] studied the structure and technical characteristics of PC and an embedded video server, analyzed the key technology of the video server, and prospected the development trend of the video server.

Figure 7. Schematic diagram of intelligent video surveillance technology.

5.5. Coal Mine Robot Technology

At present, underground coal mine robots mainly include coal mine intelligent rapid tunneling robots, intelligent coal mining robots in coal mines, coal mine intelligent transportation robots, coal mine intelligent inspection robots and coal mine intelligent detection and disaster relief robots.

The research of coal mine detection and disaster relief robots mainly focuses on power system, explosion-proof design, motion control, and information monitoring and transmission. The inspection robot should have basic functions such as video intelligent monitoring, data transmission and analysis. Inspection robots are divided into belt conveyor inspection robots and fully-mechanized mining face inspection robots. The main functions of the belt conveyor inspection robot include autonomous cruise, autonomous positioning, autonomous obstacle avoidance, autonomous charging and autonomous dust removal. The key to the inspection robot for fully-mechanized mining face [52] mainly includes flexible track technology, precise positioning and navigation technology, precise control technology, dynamic image acquisition technology and 3D stope model construction technology. Song et al. [53] reviewed the application status of five types of coal mine robots in China and abroad, including tunneling, coal mining, transportation, safety control and rescue, studied the application status of bionic robot technology in coal mine operations in China and abroad, and pointed out the development ideas and research directions of coal mine bionic robot technology and equipment. Intelligent coal mining robot in coal mines refers to the precise control of coal face shearers, scraper conveyors, hydraulic supports, transfer machines and advanced supports through integrated intelligent control systems, autonomous operation and multi-machine coordinated linkage operations.

Figure 8 shows the working schematic diagram of the underground coal mine inspection robot. On 23 June 2020, the coal mine intelligent rapid excavation robot system jointly developed by Shanxi Coal Yubei Coal Industry Co., Ltd., Xi'an University of Science and Technology and Xi'an Coal Mining Machinery Co., Ltd. was officially released. On 10 September, the first domestic coal mine intelligent rapid tunneling robot system with a total length of about 100 m, a height of 4.5 m, and a weight of 400 tons was put into use in Shanxi Coal's Yubei Coal Industry.

Figure 8. Working diagram of the underground inspection robot in coal mine.

6. Intelligent Unmanned Mining Technology Mode

6.1. Fully-Mechanized Mining Equipment Automation and Remote Visualized Intervention

The main technical support of the intelligent unmanned mining technology mode of "fully-mechanized mining equipment automation and remote visual intervention" at the working face is the automation of a complete set of fully-mechanized mining equipment [54]. Technicians observe and analyze the data collected by sensor equipment, and realize unmanned mining at the working face based on the automation of fully-mechanized mining equipment and manual remote visual intervention. Most of the existing intelligent mining working face in China adopt this kind of intelligent unmanned mining technology mode.

The automation of comprehensive mining equipment is mainly through the integrated control system software to control the main control computer software of the electro-hydraulic control system, the main control computer software of the integrated liquid supply system, the main control computer software for the centralized control of the slot, the industrial Ethernet network management software, the video management software, and the data integrated software, data communication software, etc., are integrated on a unified platform, and run on multiple explosion-proof computer hardware platforms at the same time, to realize distributed integrated control, and complete the fully-mechanized mining equipment of the fully-mechanized mining face, including hydraulic supports, shearers and belts, centralized monitoring and control of machines, scrapers, transfer machines, crushers, belts, pumping stations and other equipment [55].

Remote visualization intervention is mainly through video visualization technology, remote real-time control technology, automatic data push technology, etc., to achieve remote monitoring and control of comprehensive mining equipment in the monitoring center.

6.2. Intelligent Adaptive Mining Technology Mode

In recent years, with the development of new technologies such as the Internet of Things, 5G, and artificial intelligence and their application in the coal industry, a higher level of intelligent unmanned coal mining mode has gradually formed an intelligent adaptive mining technology mode [56]. The intelligent adaptive mining technology mode is based on "integrated mining equipment automation and remote visualization intervention", making full use of machine vision, multi-source information fusion and three-dimensional physical simulation technology to achieve intelligent analysis of collected data, and automatically make it based on the analysis results; for example, the self-adaptive height adjustment, self-adaptive adjustment and self-adaptation of the shearer drum, so as to truly realize the small and unmanned mining in the coal industry [57].

On 12 September 2020, the "Transparent Intelligent Fully-Mechanized Face Adaptive Coal Mining Key Technology and System" project of Guotun Coal Mine of the Linkuang Group passed the appraisal of scientific and technological achievements by the Appraisal Committee. It is reported that the project has achieved the first time in many coal industries, including the first application of 5G technology to the normal production of underground intelligent and adaptive fully-mechanized mining faces, the first development of a TGIS management and control platform including a digital twin system, and the first development of a full-scale control platform. The dynamic and precise positioning system of the automatic measuring robot makes the intelligent unmanned mining technology move from memory cutting to intelligent adaptive cutting, laying a solid high-tech foundation for the development of intelligent unmanned mining technology, and the social and economic benefits are significant.

6.3. Theory and Technology of Intelligent Unmanned Mining Overall Design System

Figure 9 shows the overall design framework system diagram of intelligent unmanned mining. A unified 4DGIS, virtual mine and configuration software platform was used to manage the spatial data and attribute data of the entire safety production process of "mining, excavation, machine, transportation, and communication", using a unified GIS, three-dimensional visualization or virtual mine platform [58]. For the integrated automation system, a unified configuration software platform is adopted; production mine operation management, safety production online inspection management, safety production technology comprehensive management, and decision support adopts a unified management platform to achieve the height of the software and hardware system integrated operation, analysis and management; unified data transmission, underground enterprise management, integrated automation, online inspection, and integrated management of safe production uses a unified network for transmission; unified data warehouse, production mine operation management, integrated automation, safe production online inspection management, comprehensive management of safety production technology, and decision support adopt a unified data warehouse to realize data sharing [59–61].

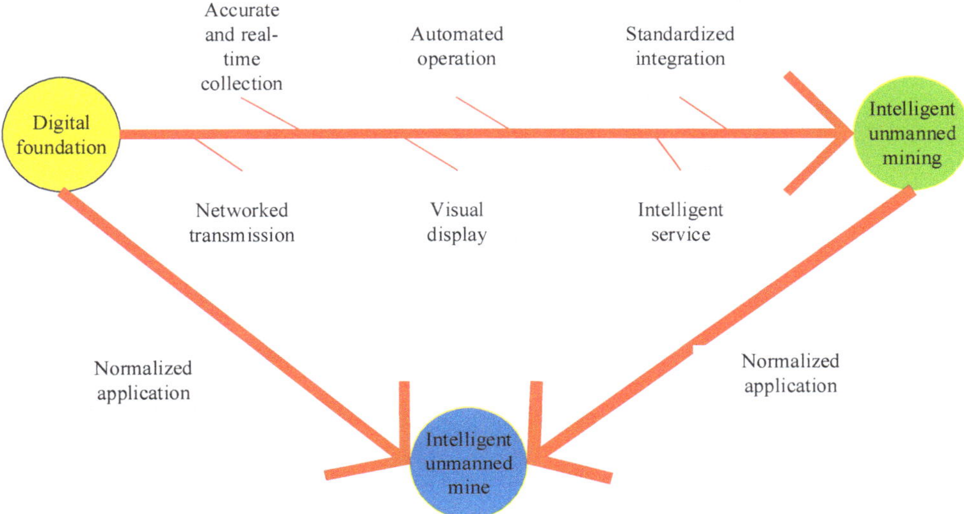

Figure 9. System diagram of the overall design architecture of intelligent unmanned mining.

7. Problems with Intelligent Unmanned Mining

(1) Intelligent fully-mechanized mining equipment. Although China has realized the localization of most of the fully-mechanized mining equipment in the intelligent working face, the intelligent fully-mechanized mining equipment still has problems such as poor reliability, low perception accuracy and poor coordination. At the same time, the sensor technology of fully-mechanized mining equipment also needs further development. Although new optical fiber sensors and MEMS sensors [62] have higher measurement accuracy, are lower in cost, have more functions and easier realization of intelligence than traditional sensors, they are limited by the complex underground environment of coal mines, and multiple functions are still unable to be realized.

(2) The key technology of intelligent unmanned mining needs further development. This includes large-section roadway deformation intelligent control technology, intelligent technology of roadheaders, intelligent technology of bolt support [63], intelligent technology of transportation system, intelligent technology of video monitoring, inertial navigation technology [64] and geological exploration technology, etc.

(3) The degree of intelligence is low. Due to the limitation of technology and equipment, the intellectualization degree of unmanned mining technology in coal mines is relatively low at present, which is mainly due to the contradiction between the static control system and the dynamic application environment.

(4) Government policy. The current relevant laws and policies are mostly instructive policies and are not compulsory. There are many local policies and the differences between different provinces are large. Different local policies have led to the uneven development of coal mine intelligence in each province and city in China. The research hotspots of coal mine intelligence are different, which is not conducive to the healthy development of China's coal industry.

(5) Talent reserve. At present, there are very few schools in Chinese universities that offer smart mining, there is lack of relevant teaching materials, and a lack of a comprehensive new talent training model.

8. Prospects for the Development of Intelligent Unmanned Mining

(1) Intelligent automation technology. Intelligent automation technology, including shearer memory cutting technology and intelligent automatic rapid tunneling technology, needs further innovation and development. It is necessary to increase investment and research on intelligent mining technology, and promote the development of shearers in the direction of intelligent perception, intelligent planning and autonomous cutting. Moreover, there is a need to increase the research and development of the full-section rectangular rapid roadheader, and improve the technical level of intelligent automatic rapid roadheading in coal mines.

(2) Intelligent control technology. Through the application of 5G, inertial navigation, gigabit industrial Ethernet technology and monitoring technology in the key technologies of intelligent unmanned mining in coal mines, the operation accuracy and application distance of intelligent control are improved, and the reaction time and control error of intelligent control are reduced. Intelligent control technology that realizes fast control and precise control is required

(3) Intelligent monitoring technology. Through continuous improvement of communication quality, data transmission speed, camera resolution and monitoring system stability, in-depth study of high-definition imaging of intelligent working faces based on thermal imaging technology can be conducted, to achieve comprehensive monitoring of coal mining machines, scraper conveyors, hydraulic supports, etc., for real-time tracking of mining equipment.

(4) Other smart technologies. The existing underground coal mine robots are still unable to perform complex underground operations due to technological and technical limitations. In the future, coal mine robots should have multi-joint manipulators like industrial robots, which can achieve multiple degrees of freedom and can complete various complex underground operations. At the same time, we should vigorously develop coal mine intelligent rapid excavation robots and coal mine intelligent detection and disaster relief robots.

9. Conclusions

The basic concepts and characteristics of intelligent unmanned mining are introduced, and the related research that may be beneficial to this research content is reviewed from four aspects of basic theory, key technology, mining mode, and overall design system theory and technology. Finally, the research status and future development trend of intelligent unmanned mining are summarized.

This paper discussed and analyzed intelligent unmanned mining from the relationship between key technologies and engineering applications, as the focus on 5G + intelligent mines and intelligent equipment. This paper introduced the basic concepts and characteristics of intelligent unmanned mining. Then we built a research structure of technology-driven mode, in which four layers were described in detail. Finally, the existing problems and development direction were put forward.

(1) By studying the development history of my country's intelligent unmanned mining technology, the definition of intelligent unmanned mining is given. It introduces, in detail, the fundamental role of multi-source heterogeneous data models and geological prediction models in the field of intelligent unmanned mining.

(2) Some key technologies of intelligent unmanned mining are introduced, and three major problems in intelligent unmanned mining are given. Including the poor reliability of intelligent fully-mechanized mining equipment, low perception accuracy, and poor coordination, the key technologies of intelligent unmanned mining need to be further developed and the degree of intelligence of unmanned mining technology in coal mines is relatively low.

(3) Aiming at the deficiencies of the key technologies of intelligent unmanned systems, the intelligent automation technology, intelligent control technology, intelligent moni-

toring technology and other intelligent technologies are prospected, and it is hoped that China's intelligent unmanned mining technology will flourish.

Author Contributions: Conceptualization, K.Z.; methodology, M.H.; software, C.Z.; validation, D.L.; formal analysis, X.C.; investigation, L.K.; resources, K.Z.; data curation, L.K.; writing—original draft preparation, L.K.; writing—review and editing, K.Z. and L.K.; supervision, X.C. and M.H.; project administration, K.Z.; funding acquisition, K.Z. All authors have read and agreed to the published version of the manuscript.

Funding: This research was funded by the National Natural Science Foundation of China (51804160); Funded Projects for Fundamental Scientific Research Funds of Central Universities (3142019009); Projects Supported by the Natural Science Foundation of Hebei Province (E2019508209); and the State Key Laboratory of Deep Geomechanics and Underground Engineering (Beijing) Open Fund Project (SKLGDUEK1822, SKLGDUEK2133).

Data Availability Statement: The data used to support the findings of this study are available from the corresponding author upon request.

Conflicts of Interest: The authors declare that there is no conflict of interest regarding the publication of this paper.

References

1. Wang, G.F.; Pang, Y.H.; Ren, H.W. Intelligent coal mining pattern and technological path. *J. Min. Strat. Control. Eng.* **2020**, *2*, 5–19.
2. Ge, S.R.; Hao, S.Q.; Zhang, S.H.; Zhang, X.F.; Zhang, L.; Wang, S.B.; Wang, Z.B.; Bao, J.S.; Yang, X.L.; Yang, J.J. Status of intelligent coal mining technology and potential key technologies in China. *Coal Sci. Technol.* **2020**, *48*, 28–46.
3. Fan, J.D.; Yan, Z.G.; Li, C. Exploration of intelligent coal mining key technology based on 5G technology. *Coal Sci. Technol.* **2020**, *48*, 92–97.
4. Guidance on Accelerating the Intelligent Development of Coal Mines. *China Coal Daily*, 5 March 2020.
5. Wang, G.F.; Zhao, G.R.; Ren, H.W. Analysis of key core technologies of intelligent coal mining and intelligent mining. *Coal J.* **2019**, *44*, 34–41.
6. Guo, C.; Yang, Z.; Chang, S.; Ren, T.; Yao, W. Precise Identification of Coal Thickness by Channel Wave Based on a Hybrid Algorithm. *Appl. Sci.* **2019**, *9*, 1493. [CrossRef]
7. Wang, T.; Jie, J.; Lin, Z.Y.; Fang, H.M.; Wang, Y.; Liu, Y.F. Coordinated Exploration Model and its Application to Coal and Coal-associated Deposits in Coal Basins of China. *Acta Geol. Sin.* **2021**, *95*, 1346–1356. [CrossRef]
8. Hao, Y.; Wu, Y.; Ranjith, P.G.; Zhang, K.; Zhang, H.Q.; Chen, Y.L.; Li, M.; Li, P. New insights on ground control in intelligent mining with Internet of Things. *Comput. Commun.* **2020**, *150*, 788–798. [CrossRef]
9. Li, J.; Zhan, K. Intelligent Mining Technology for an Underground Metal Mine Based on Unmanned Equipment. *Engineering* **2018**, *4*, 381–391. [CrossRef]
10. Wang, G.F.; Xu, Y.X.; Ren, H.W. Intelligent and ecological coal mining as well as clean utilization technology in China: Review and prospects. *Int. J. Min. Sci. Technol.* **2019**, *29*, 161–169. [CrossRef]
11. Wang, G.F.; Liu, F.; Pang, Y.H.; Ren, H.W.; Ma, Y. Intelligent coal mine—Core technology support for high-quality development of coal industry. *J. Coal* **2019**, *44*, 349–357. [CrossRef]
12. Xie, H.P.; Wang, J.H.; Wang, G.F.; Ren, H.W.; Liu, J.Z.; Ge, S.R.; Zhou, H.W.; Wu, G.; Ren, S.H. The new concept of coal revolution and the concept of coal science and technology development. *J. Coal* **2018**, *43*, 1187–1197.
13. Cheng, G.M.; Huang, K.; Wang, S.J. Spiral drilling technology and its development. *Min. Metall. Eng.* **2003**, *23*, 4–6.
14. Zhao, L.J.; He, J.Q.; Li, F.Q. Numerical simulation of coal plow cutting process. *J. China Coal Soc.* **2012**, *37*, 878–883.
15. Basarir, H.; Oge, I.F.; Aydin, O. Prediction of the stresses around main and tail gates during top coal caving by 3D numerical analysis. *Int. J. Rock Mech. Min. Sci.* **2015**, *76*, 88–97. [CrossRef]
16. Massinaei, M.; Jahedsaravani, A.; Taheri, E.; Khalilpour, J. Machine vision based monitoring and analysis of a coal column flotation circuit. *Powder Technol.* **2019**, *343*, 330–341. [CrossRef]
17. Wang, J.; Huang, Z. The recent technological development of intelligent mining in China. *Engineering* **2017**, *3*, 439–444. [CrossRef]
18. Li, L.G.; Sun, D.Y.; Han, G.G.; Li, X.B.; Hu, Q.C.; Shu, L. Velocity-free Localization of Autonomous Driverless Vehicles in Underground Intelligent Mines. *IEEE Trans. Veh. Technol.* **2020**, *69*, 9292–9303.
19. Zhang, K.X.; Li, S.B.; He, M.C.; Ning, Y.; Zhang, L.; Huang, Z.H. Study on key technologies of intelligent unmanned coal mining series I: Study on diagonal adjustment control technology of intelligent fully—Mechanized coal mining face. *Coal Sci. Technol.* **2018**, *46*, 139–149.
20. Wang, G.F.; Wang, H.; Ren, H.W.; Zhao, G.R.; Pang, Y.H.; Du, Y.B.; Zhang, J.H.; Hou, G. The 2025 scenario goal and development path of smart coal mine. *Coal J.* **2018**, *43*, 295–305.
21. Wang, G.F.; Du, Y.B. Development direction of intelligent coal mine and intelligent mining technology. *Coal Sci. Technol.* **2019**, *47*, 1–10.

22. Li, S.B. Progress and development trend of intelligent mining technology. *Coal Sci. Technol.* **2019**, *47*, 102–110. [CrossRef]

23. Fan, J.D. Research on key technologies of intelligent fully mechanized mining on working face with large mining height. *Ind. Mine Autom.* **2018**, *44*, 363–371.

24. Zhang, K.X. Study on intelligent mining technology of fully-mechanized heading face. *Coal Sci. Technol.* **2017**, *45*, 106–111.

25. Zhang, K.X.; Wang, X.L.; He, M.C.; Yin, S.X.; Li, S.B.; Sun, J.D.; Li, D.; Cheng, Z.H.; Zhao, Q.F.; Yin, S.F.; et al. Research on multi-level fuzzy comprehensive evaluation of the applicability of intelligent unmanned mining face. *J. Min. Strat. Control. Eng.* **2021**, *3*, 47–56.

26. Huang, Z.H.; Wang, F.; Zhang, S.X. Research on the architecture and key technologies of intelligent coal mining system. *J. China Coal Soc.* **2020**, *45*, 1959–1972.

27. Wang, C.F.; Rong, Y. Concept, architecture and key technologies for transparent longwall face. *Coal Sci. Technol.* **2019**, *47*, 156–163. [CrossRef]

28. Wang, J.H.; Huang, L.T.; Li, S.B.; Huang, Z.H. Development of intelligent technology and equipment in fully-mechanized coal mining face. *J. China Coal Soc.* **2014**, *39*, 1418–1423.

29. Yuan, J.H.; Luo, X.G.; Li, Y.; Hu, X.Q.; Chen, W.C.; Zhang, Y. Multi criteria decision-making for distributed energy system based on multi-source heterogeneous data. *Energy* **2022**, *239*, 122250. [CrossRef]

30. Yuan, J.H.; Luo, X.G.; Li, Z.D.; Li, L.F.; Ji, P.L.; Zhou, Q.; Zhang, Z.L. Sustainable development evaluation on wind power compressed air energy storage projects based on multi-source heterogeneous data. *Renew. Energy* **2021**, *169*, 1175–1189. [CrossRef]

31. Zhang, K.X.; Xu, L.X.; Li, X.; Mao, M.C.; Fu, D.L.; Zhang, Y.L.; Kang, L.; Wang, X.L. Research on big data analysis and decision-making method and system of intelligent mining in transparent working face. *Coal Sci. Technol.* **2021**. Available online: https://kns.cnki.net/kcms/detail/detail.aspx?dbcode=CAPJ&dbname=CAPJLAST&filename=MTKJ20211202 00&uniplatform=NZKPT&v=hhPV5LJlyhYeSh63RHRJvvfZbtlJmfedlYCjk6UnM0DXLk34DviH7kQrFQHuadZ4 (accessed on 5 December 2021).

32. Gao, M.; Hao, H.; Xue, S.; Lu, T.; Cui, P.; Gao, Y.; Xie, J.; Yang, B.; Xie, H. Discing behavior and mechanism of cores extracted from Songke-2 well at depths below 4500 m. *Int. J. Rock Mech. Min. Sci.* **2022**, *149*, 104976.

33. Gao, M.; Xie, J.; Gao, Y.; Wang, W.; Li, C.; Yang, B.; Liu, J.; Xie, H. Mechanical behavior of coal under different mining rates: A case study from laboratory experiments to field testing. *Int. J. Min. Sci. Technol.* **2021**, *31*, 825–841. [CrossRef]

34. Gao, M.Z.; Zhang, Z.L.; Yin, X.G.; Xu, C.; Liu, Q.; Chen, H.L. The location optimum and permeability-enhancing effect of a low-level shield rock roadway. *Rock Mech. Rock Eng.* **2018**, *51*, 2935–2948. [CrossRef]

35. Notice on Soliciting Opinions on 4 China Coal Industry Association Group Standards including "Code for Design of Coal Mining Faces Based on Dynamic Geological Model and Independent Planning Mining". *China Coal Assoc. Sci. Technol. Lett.* **2021**. Available online: http://www.coalchina.org.cn/index.php?m=content&c=index&a=show&catid=61&id=132805 (accessed on 5 December 2021).

36. Dong, L.G.; Tong, X.J.; Ma, J. Quantitative Investigation of Tomographic Effects in Abnormal Regions of Complex Structures. *Engineering* **2020**, *7*, 1011–1022. [CrossRef]

37. Dong, L.G.; Tong, X.J.; Hu, Q.C.; Tao, Q. Empty region identification method and experimental verification for the two-dimensional complex structure. *Int. J. Rock Mech. Min. Sci.* **2021**, *147*, 104885. [CrossRef]

38. Dong, L.G.; Hu, Q.C.; Tong, X.J.; Liu, Y.F. Velocity-Free MS/AE Source Location Method for Three-Dimensional Hole-Containing Structures. *Engineering* **2020**, *6*, 827–834. [CrossRef]

39. Cao, B.F.; Luo, X.R.; Zhang, L.K.; Lei, Y.H.; Zhou, J.S. Petrofacies prediction and 3-D geological model in tight gas sandstone reservoirs by integration of well logs and geostatistical modeling. *Mar. Pet. Geol.* **2020**, *114*, 104202. [CrossRef]

40. Zhang, K.X.; Yang, H.J.; He, M.C.; Sun, J.D.; Li, D.; Cheng, Z.H.; Zhao, Q.F.; Yin, S.X.; Kang, L.; Wang, X.L.; et al. Comprehensive evaluation of intelligent fully mechanized mining face based on grey relational analysis. *Mod. Tunn. Technol.* **2021**. Available online: http://kns.cnki.net/kcms/detail/51.1600.U.20210831.1317.004.html (accessed on 5 December 2021).

41. Li, S.B. Present situation and prospect on intelligent unmanned mining at work face. *China Coal* **2019**, *45*, 5–12.

42. Liu, C.S.; Chen, J.G. Mathematic Model of Memory Cutting for Coal Shearer Based on Single Demo Knife. *Coal Sci. Technol.* **2011**, *39*, 71–73.

43. Fan, Q.G.; Li, W.; Wang, Y.Q.; Fan, M.B.; Yang, X.F. A memory cutting algorithm for shearer using grey Markov combination model. *J. Cent. South Univ.* **2011**, *42*, 3054–3058.

44. Zhang, L.L.; Tan, C.; Wang, Z.B.; Yang, X.F.; Mi, J.P. Memory cutting path optimization of shearer based on genetic algorithm. *Coal Eng.* **2011**, *23*, 111–113.

45. Zhang, L.L.; Tan, C.; Wang, Z.B.; Mi, J.P.; Zhu, W.C. Mining machine memory cutting path optimization based on particle swarm optimization algorithm. *Coal Sci. Technol.* **2010**, *38*, 69–71.

46. Ren, H.W.; Meng, X.G.; Li, Z.; Li, M.Z. Study on key technology of intelligent control system applied in 8 m large mining height fully-mechanized face. *Coal Sci. Technol.* **2017**, *45*, 37–44.

47. Huang, Z.H.; Miao, J.J. The application research of equipment centralized control technology in fully mechanized mining face. *Coal Sci. Technol.* **2013**, *41*, 14–17.

48. Song, D.Y.; Song, J.C.; Tian, M.Q.; Xu, C.Y.; Song, X.; Li, X.S. Development and application of electro-hydraulic control technology of hydraulic support in fully mechanized coal mining face. *J. Taiyuan Univ. Technol.* **2018**, *49*, 240–251.

49. Liu, Z.R.; Liu, J. The application of automation technology in fully mechanized mining face. *Inn. Mong. Coal Econ.* **2018**, *265*, 7–12.

50. Chen, G.P. The application of intelligent analysis and early warning technology in video surveillance system. *Coal Mine Mach.* **2014**, *35*, 172–173.
51. Cheng, D.Q.; Qian, J.S. Digital video surveillance server and its key technologies. *Coal Sci. Technol.* **2004**, *32*, 43–46.
52. Yang, X.J.; Wang, R.F.; Wang, H.F.; Yang, Y.K. A novel method for measuring pose of hydraulic supports relative to inspection robot using LiDAR. *Measurement* **2020**, *154*, 107452. [CrossRef]
53. Song, R.; Zheng, Y.K.; Liu, Y.X.; Ma, X.; Li, Y.B. Application and prospect analysis of bionic robot technology in coal mine. *J. Coal* **2020**, *45*, 2155–2169.
54. Huang, L.T.; Huang, Z.H.; Zhang, K.X. Key Technology of Mining in Intelligent Fully Mechanized Coal Mining Face with Large Mining Height. *Coal Min.* **2016**, *21*, 1–6.
55. Huang, Z.H.; Nan, T.F.; Zhang, K.X.; Feng, Y.H. Design on intelligent control platform of mechanized mining robot based on Ethernet/IP. *Coal Sci. Technol.* **2017**, *45*, 9–15.
56. Tian, C.J. Research of intelligentized coal mining mode and key technologies. *Ind. Mine Autom.* **2016**, *42*, 28–32.
57. Wang, G.F.; Du, Y.B.; Ren, H.W.; Fan, J.D.; Wu, Q.Y. Top level design and practice of smart coal mines. *J. China Coal Soc.* **2020**, *45*, 1909–1924.
58. Li, C.H.; Mao, S.J.; Jiang, Y.F. Design and Implementation of Mine Virtual Platform and Electromechanical Monitoring and Measuring Application. *Coal Sci. Technol.* **2011**, *39*, 94–99.
59. Zoubir, O.; Doulkifli, B.; Omar, B.; Rachid, C. AuMixDw: Towards an automated hybrid approach for building XML data warehouses. *Data Knowl. Eng.* **2019**, *120*, 60–82.
60. Zeng, L.Y.; Liu, S.Q.; Erhan, K.; Paul, C.; Mahmoud, M. A comprehensive interdisciplinary review of mine supply chain management. *Resour. Policy* **2021**, *74*, 102274. [CrossRef]
61. Song, Z.Q.; Chen, L.L.; Wang, C.Q.; Liu, X.G. Understanding of safe mining conditions for fully mechanized top coal caving. *J. China Coal Soc.* **1995**, *18*, 356–360.
62. Rao, K.; Liu, H.F.; Wei, X.L.; Wu, W.J.; Hu, C.Y.; Fan, J.; Liu, J.Q.; Tu, L.C. A high-resolution area-change-based capacitive MEMS tilt sensor. *Sens. Actuators A Phys.* **2020**, *313*, 112191. [CrossRef]
63. Gao, M.Z.; Xie, J.; Guo, J.; Lu, Y.Q.; He, Z.Q.; Li, C. Fractal evolution and connectivity characteristics of mining-induced crack networks in coal masses at different depths. *Geomech. Geophys. Geo-Energy Geo-Resour.* **2021**, *7*, 9. [CrossRef]
64. Wang, Z.H.; Cheng, X.H. Adaptive Optimization Online IMU Self-Calibration Method for Visual-Inertial Navigation System. *Measurement* **2021**, *180*, 109478. [CrossRef]

Article

Wave Velocity in Sandstone and Mudstone under High Temperature and Overpressure in Yinggehai Basin

Zichun Liu *, Xiangdong Du, Zhenyu Zhu and Xin Li

Exploration and Development Research Department, CNOOC Research Institute Ltd., Beijing 100028, China; duxd@cnooc.com.cn (X.D.); zhuzhy2@cnooc.com.cn (Z.Z.); lixin11@cnooc.com.cn (X.L.)
* Correspondence: liuzch7@cnooc.com.cn

Abstract: Wave velocity under different pressure and temperature (PT) conditions plays an important role in the exploration of oil and gas reservoirs. We obtained the mineral composition and porosity of 20 underground sandstone and mudstone samples in Yinggehai Basin via X-ray diffraction and porosity measurements. Using high-frequency ultrasound, the P- and S-wave velocities of four samples under high temperature and overpressure conditions were found to vary significantly, owing to the material composition and porosity. According to the comparison between the experimental conclusion and the well-logging data, the genesis of false bright spot and dark spot gas reservoirs in the study area was analyzed. The variation in P-wave velocity under different temperature and pressure conditions was explained with the PT coefficient. The traditional pressure–velocity and temperature–velocity prediction methods were improved and applied to well-logging data. Herein, the velocity of P- or S-waves of sand and mudstone under high temperature and overpressure via rock physics experiments and the genesis of false bright spot and dark spot gas reservoirs in the Yinggehai Basin was observed. Overall, the results serve as a theoretical basis for seismic exploration in the study area.

Keywords: seismic rock physics; sandstone and mudstone; P-wave and S-wave velocity; pressure; temperature

Citation: Liu, Z.; Du, X.; Zhu, Z.; Li, X. Wave Velocity in Sandstone and Mudstone under High Temperature and Overpressure in Yinggehai Basin. *Energies* **2022**, *15*, 2615. https://doi.org/10.3390/en15072615

Academic Editors: Gan Feng, Qingxiang Meng, Fei Wu, Gan Li and Manoj Khandelwal

Received: 9 February 2022
Accepted: 10 March 2022
Published: 3 April 2022

Publisher's Note: MDPI stays neutral with regard to jurisdictional claims in published maps and institutional affiliations.

1. Introduction

The Yinggehai Basin is a typical high-temperature and overpressure basin located in the northern sea area of the South China Sea. Since the 1990s, many normal pressure and temperature (PT) gas fields have been found in the shallow layer of the central diapir belt using conventional bright spot technology [1]. However, as the exploration involved high temperature and overpressure, the bright spot technology adopted for the middle and deep layers of Dongfang and Ledong areas failed in the late 1990s [2]. Relevant research data indicate that the physical properties and seismic response of middle and deep reservoirs in this area are highly complex owing to the influence of high temperature and overpressure. Further, abnormal changes in P-wave velocity and impedance have been found, resulting in false bright or dark spots in seismic data (Figure 1) [3,4]. Owing to this phenomenon, traditional reservoir prediction techniques, such as conventional bright spots and pre-stack wave impedance inversion, were challenging to employ in the study area. At present, there is no consensus on the rock wave velocity anomaly and the influencing factors under high temperature and overpressure in this area. Accordingly, a comprehensive rock physics investigation is needed.

The PT of the rock had a significant impact on the wave velocity. The relationship between PT and velocity is caused by continuous pore closure and phase transformation of minerals in rocks [5,6]. Based on laboratory results, P- and S-wave velocity are usually logarithmically positively correlated with confining pressure and linearly negatively correlated with temperature and pore pressure [7–9]. Laboratory measurement results also enabled some scholars to establish a comprehensive fitting formula between P-wave velocity and PT [10,11]. However, P- and S-wave velocity has different rates of change for similar rocks under different PT conditions [8,9]. As the factors influencing the PT coefficient in the fitting formula established in previous studies have not been discussed, the difference in P- and S-wave velocity variation of the rock cannot be explained. Combined with laboratory measurement results, this study sought to connect the influencing factors with the correlation coefficient, establish velocity–temperature and velocity–pressure relations for sandstone and mudstone, and applied these relations to the logging data.

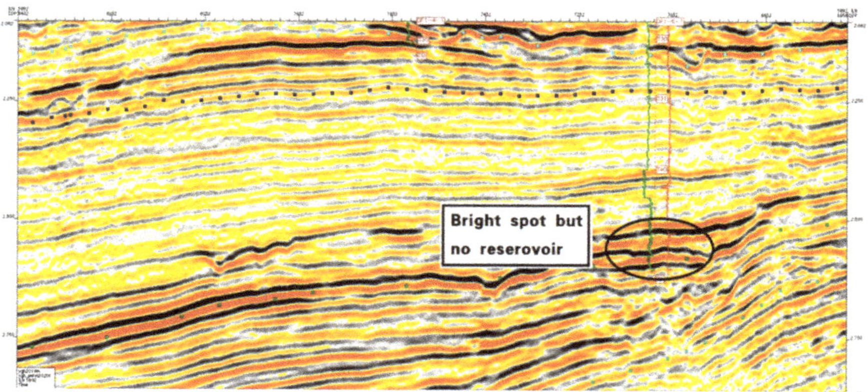

Figure 1. Seismic profile through the DF1-X6 well.

The controlling and influencing factors of wave velocity at high temperatures and overpressure formation in the Yinggehai Basin remain ambiguous. In this study, the differences in P- and S-wave velocity under high temperature and overpressure of sandstone and mudstone core samples in the Yingqiong Basin were analyzed by systematic experiments from the perspectives of minerals and pores. The following specific experimental workflow was employed: (1) the mineral composition of the samples was obtained by X-ray diffraction measurement; (2) the porosity of samples under different confining pressure was obtained by the helium method, (3) the relevant sample sections were observed; (4) the ultrasonic velocities of four samples were measured under different PT and analyzed, and (5) the fitting relationship between P-wave velocity and PT was established to predict the actual formation data.

2. Experimental Conditions and Basic Petrophysical Characteristics

Samples were collected from 20 downhole cores (2000–4000 m) in Dongfang and Ledong districts of the Yingqiong Basin, South China Sea. The mineral composition and porosity were obtained using X-ray diffraction analysis and helium experiments. The cores were processed by instrument to a diameter of 25.4 mm and a height of 40–60 mm plugs. The two end faces of the plug were finely polished to a degree less than 0.1 mm. Due to the softening effect of clay dehydration on the rock skeleton, we referred to the previous drying methods, stored the cores in an 80 °C oven for 72 h, and placed the samples indoors for 48 h to obtain samples containing 2–3% water to eliminate the softening effect of water on the rock skeleton [12,13]. The samples were analyzed by changing alone the confining pressure, temperature, or pore pressure to avoid any interference from different influencing factors on the analysis results. Autolab2000c equipment in CNOOC Research Institute produced by NER Company (Yantai, China) was used to measure the high-frequency P- and S-wave velocity. Figure 2 shows the AutoLab system and scheme of preparation for wave velocity measurements in samples. P-wave and S-wave velocity according to the direction of vibration and propagation were evaluated, namely $V_{P-0°}$ and $V_{S-0°}$ (sample cut normal to bedding), where $V_{SH-0°} \approx V_{SV-0°}$. The V_P and V_S data from cited for comparison in the paper all in this direction.

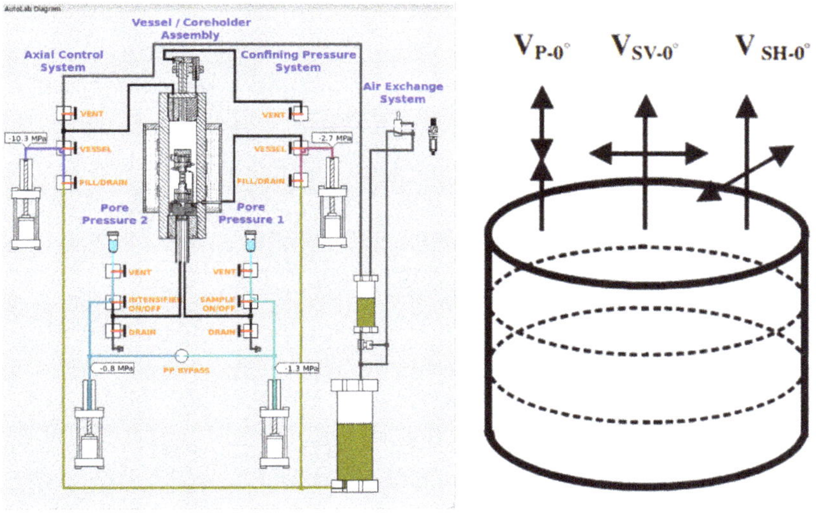

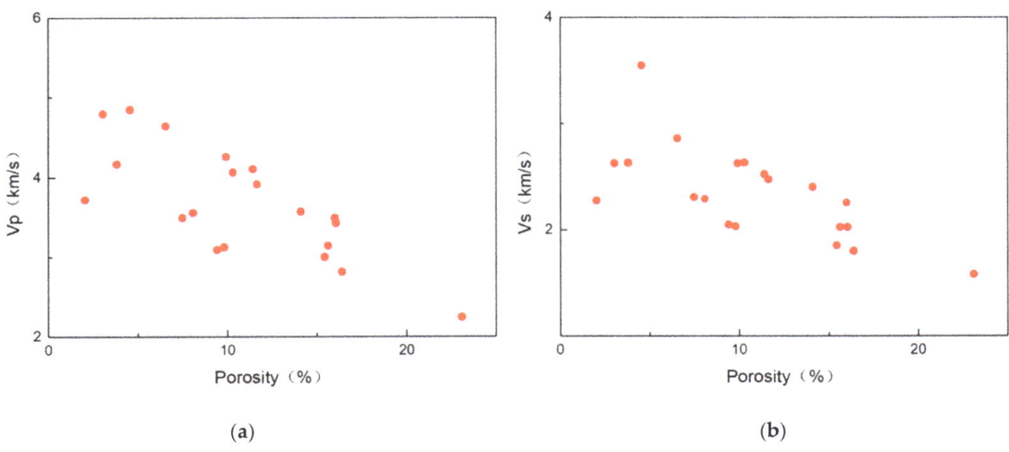

(a)   (b)

Figure 2. (**a**) AutoLab system and (**b**) scheme of preparation and P- and S-wave velocity measurements in samples [14].

According to some studies, porosity has important influence on the wave velocity of rocks [15–18]. The data depicted in Figure 3 aligns with the notion that the increase of porosity can significantly reduce P- and S-wave velocity, indicating that porosity directly determines wave velocity. Wave velocity is also related to mineral composition [19–21]. The research shows (Table 1) that quartz and feldspar have higher density and elastic parameters, with more "stiffer" elastic properties, while clay minerals have more "soft" elastic properties. The differences in mineral elasticity also lead to different velocities.

(a)   (b)

Figure 3. Porosity versus (**a**) P- and (**b**) S-wave velocity.

Table 1. Elastic parameters and density of common minerals [12].

Mineral Type	Bulk Modulus (GPa)	Shear Modulus (GPa)	Vp (km/s)	Vs (km/s)	Density (g/cc)
Clay (kaolinite)	1.5	1.4	1.44	0.93	1.58
Feldspar	37.5	15	4.68	2.39	2.62
Quartz	37	44	6.05	4.09	2.65

In order to understand the comprehensive influence of pores and minerals on elastic properties, we regarded the rock as composed of quartz, feldspar, clay, and pores. Then, the dry rock was constructed by four elastic boundary theory, with Voight for upper limit (black line), Reuss for lower limit (red line), Hashin–Shtrikman for upper and lower limit (blue and pink line), and V-R-H for average of equivalent elastic modulus of the medium. The equations and details of each model are shown in the Appendix A. The processes were as follows: (1) clay and pores as soft materials mixed with the V-R-H model, (2) quartz and feldspar as stiffer minerals mixed with the V-R-H model, (3) soft and stiffer substances were mixed through Voight, Reuss, and Hashin–Shtrikman boundary theory to obtain the elastic modulus boundary of dry rocks.

As shown in Figure 4, except for single-phase minerals, isotropic mixtures of HS model could never reach the stiffness of Voight limit [12]. When the stiffer minerals increased, the elastic parameters showed a little positive trend along the model line. The elastic parameters also showed a little downward trend with an increase in soft materials. For high-temperature and overpressure reservoir, due to the difference in elastic properties between stiffer minerals and soft materials, the rock skeletons of different components that resisted PT also differed. Further, differences in the deformation of the skeleton were found, which led to different P- and S-wave velocities and impedance relationships. Thus, the study area had false bright spots in the seismic data or formed a dark spot gas reservoir. We proceeded to analyze the rock P- and S-wave velocity difference and its causes under the high temperature and high pressure to provide a theoretical basis for reservoir oil and gas prediction in the study area.

In order to prevent too many influencing factors (mineral, porosity, pressure, and temperature) from affecting the velocity, single-physical-parameter differences of four samples were selected (samples with similar porosity and large difference in mineral content or similar mineral and large difference in porosity), and single changes in the confining pressure, temperature, and pore pressure were analyzed to avoid any interference from different influencing factors on the analysis results. Table 2 shows the porosity and mineral composition of the four selected samples.

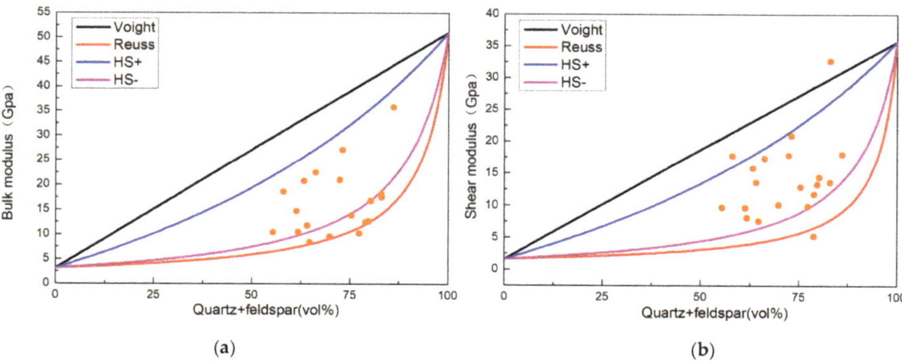

(a)   (b)

Figure 4. *Cont.*

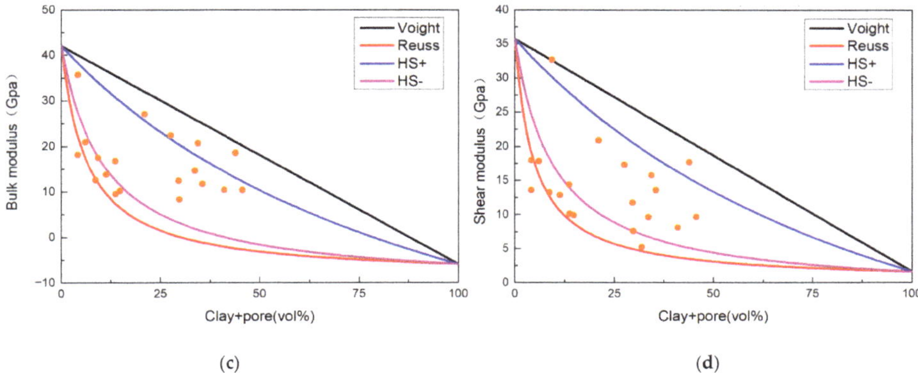

(c)   (d)

Figure 4. Effects of stiffer and soft materials on the elastic properties of samples, cross-plot of (**a**) bulk modulus and stiffer minerals, (**b**) shear modulus and stiffer minerals, (**c**) bulk modulus and soft materials, and (**d**) shear modulus and soft materials. The solid lines in the figure are the boundary of simulation (the black line is the Voight boundary, the red line is the Reuss boundary, the blue line is the upper limit of Hashin–Shtrikman model, and the pink line is the lower limit of Hashin–Shtrikman).

Table 2. Mineral content and porosity of the four samples.

Number	Quartz (%)	Potash Feldspar (%)	Plagioclase (%)	Calcite (%)	Dolomite (%)	Clay (%)	Porosity (%)
1	64.78	8.68	9.79	10.97	0.66	5.12	24.1
2	60.39	10.62	14.34	9.91	6.84	11.9	12.21
3	40.07	7.51	15.83	7.22	3.52	25.85	16.15
4	37.31	13.62	28.86	5.92	3.68	10.61	16.43

3. Influence of Pore

Porosity directly affects P- and S-wave velocity, and the resistance of different rock pores to pressure differs under high pressure. The porosity of some samples was measured by the helium method at 60 MPa (step 10 MPa). For ease of distinction, the samples were simply classified according to the porosity. As shown in Figure 5, the porosity of samples decreased with increasing pressure. However, the pore resistance of samples and the downward trend were different. Such results also led to a difference in the samples' P- and S-wave velocity under high pressure (Figure 6).

Two samples with small differences in mineral content and big differences in porosity were selected for a single pressure transformation experiment to avoid the influence of temperature and minerals on the rock. The specific parameters of these two samples are listed in Table 2. Figure 6 shows the cross-plot of P- and S-wave velocity and confining pressure under dry and completely saturated conditions (test step of 5 MPa under room temperature). The P-wave and S-wave velocities of the samples increased with the confining pressure, showing a logarithmic trend. Although the P-wave was more sensitive to pressure, all increasing trends showed two obvious differences: the P- and S-wave velocities of the high-porosity sample increased greatly with the confining pressure under low pressure (0–60 MPa) and increased slowly under high confining pressure (60–120 MPa). The P-wave velocity increased by more than 70%. Further, the wave velocity of the low-porosity sample under low confining pressure was similar to that of the high-porosity sample but hardly increased under high pressure. The P-wave velocity generally increased by <50%.

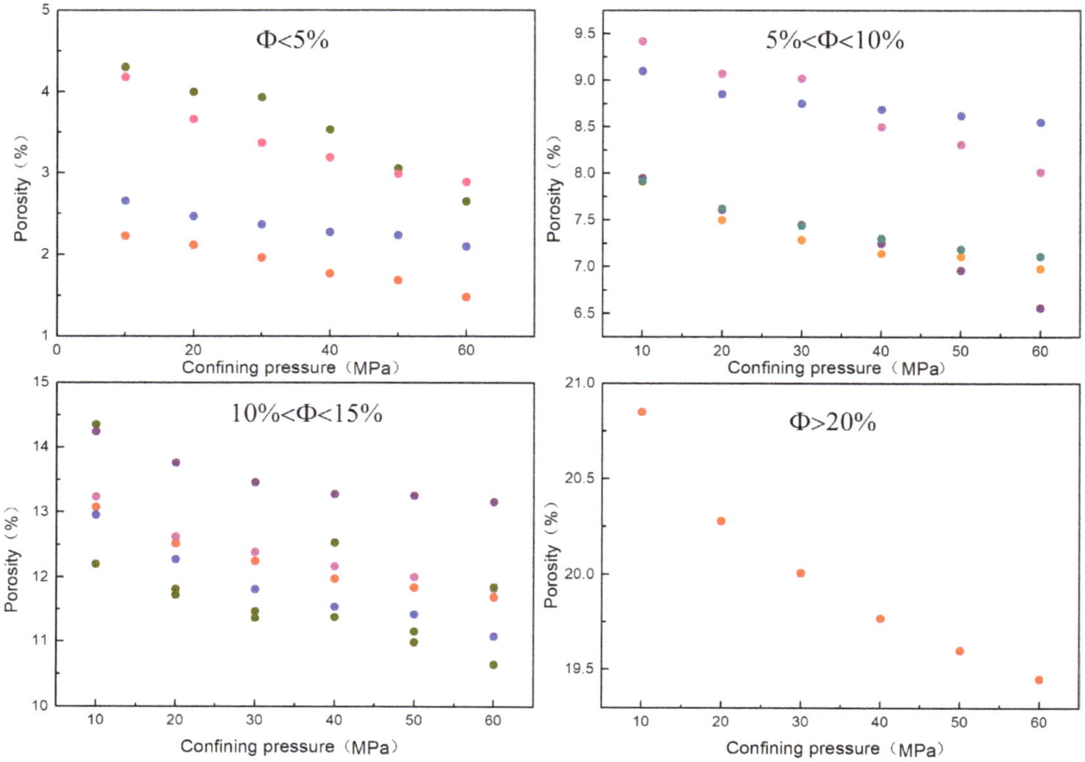

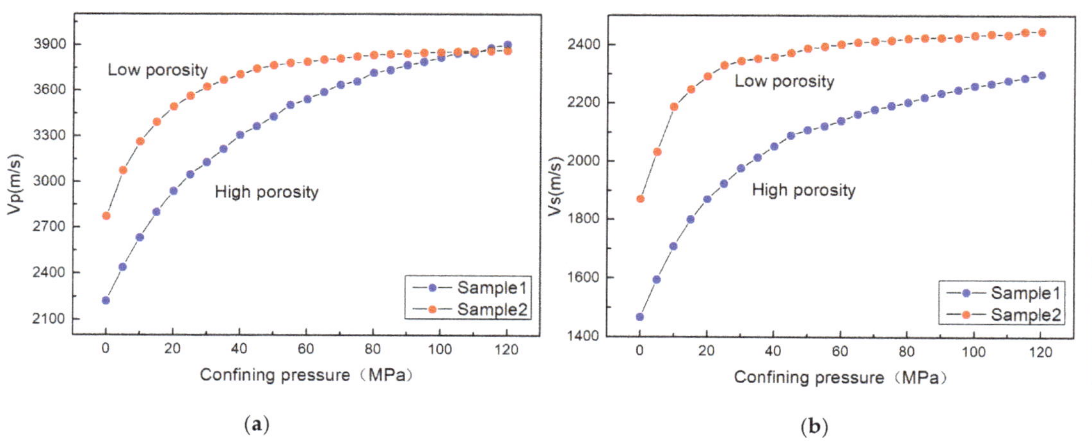

Figure 5. Cross-plot of porosity and confining pressure (different color dots represent different samples).

Figure 6. *Cont.*

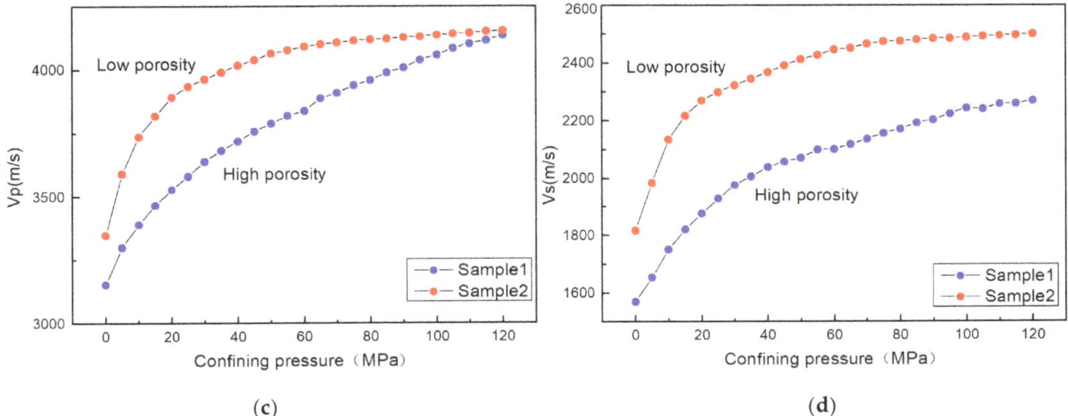

Figure 6. Relationship between P- and S-wave velocity and confining pressure, (**a**,**b**) are the relationship between wave velocity and pressure under dry conditions, and (**c**,**d**) are the relationship between wave velocity and pressure under saturated conditions.

Figure 7. The thin section images of (**a**) Sample 1 and (**b**) Sample 2.

Generally, the change in elasticity of rock under confining pressure is related to the closure of pores and plastic deformation of the mineral with different degrees of alteration [22–25]. Figure 7 shows a 200 μm cast slice. Notably, the pore structure of Sample 1 was identified to be complex. In fact, the pores with large aspect ratio and small aspect ratio were found to coexist in the rock. Several pores with small aspect ratio were also close to the increase in pressure under low confining pressure, resulting in a significant increase in P- and S-wave velocity. When the confining pressure increased, only pores with large aspect ratio constantly resisted the pressure. As a result, the relationship between P- and S-wave velocity and pressure was related to the resistance of pores with large aspect ratio to pressure, which indicated that the wave velocity increased slowly with the pressure. In contrast, Sample 2 had a lower porosity and only contained pores with a low aspect ratio. After the pressure increased to 60 MPa, the pores were completely closed, and only the rock skeleton resisted the pressure. As the P- and S-wave velocity no longer increased with pressure, this highlighted the compactness of the rock. The two trends of porosity and pressure revealed that the difference in pore structure and porosity caused a change in the relationship between P- and S-wave velocity and pressure. The P- and S-wave velocity of the higher porosity sample increased slowly with confining pressure under high conditions. However, no increase was observed for the lower porosity sample. This phenomenon may be related to the genesis of the dark point gas reservoir, that is, the P- and S-wave velocity of high-porosity gas sand stratum increased abnormally under higher pressure and caused an increase in impedance, which was similar to that of the overlying background mudstone, ultimately forming a dark point gas reservoir.

The relationship between temperature and P- and S-wave velocity of the two samples was evaluated, and a cross-plot of the wave velocity and temperature of dry samples is shown in Figure 8. The wave velocities of two samples showed a downward trend with the increase of temperature, but the trend was subtly different. The P-wave velocity of the high-porosity sample decreased by more than 7%, while that of the low-porosity sample decreased by 4%. The difference in the temperature–velocity trend reflected the degree of pore expansion during the heating process of samples. Owing to the difference in porosity, the thermal expansion of the pores of high-porosity sample was greater than that of the low-porosity sample [26,27], resulting in a greater trend of P- and S-wave velocity reduction. This phenomenon showed that the high-porosity formation had the characteristics of velocity reduction under high-temperature conditions. If there are big differences in porosity and temperature between the upper and lower mudstone formations, this may lead to the difference of wave impedance and form a dark spot gas reservoir.

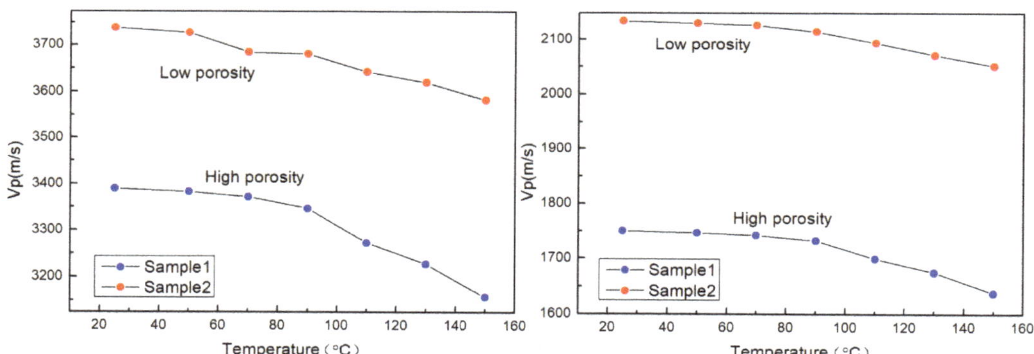

Figure 8. Cross-plot of temperature and P- and S-wave velocity.

4. Influence of Mineral Composition

The composition of the rock skeleton also influences the relationship between velocity and pressure. Sample 3 and Sample 4 with small difference in porosity and pore structure but big differences in clay mineral content were selected for the analysis. The specific parameters of these two samples are provided in Table 2. Figure 9 shows the cross-plot of the P- and S-wave velocity and confining pressure under dry and completely saturated conditions (test step of 5 MPa under room temperature). The trends of the two samples were identified to be similar under a low confining pressure, showing that the P- and S-wave velocity increased greatly with pressure. However, an obvious difference was found when pressure rose to 40–60 MPa. The P- and S-wave velocity of Sample 4 with low clay content gradually slowed down with the increase in pressure after the pressure exceeded 40–60 MPa, following which P-wave velocity increased by 16%. Sample 3 with a high clay content maintained a large range of increase, in which the P-wave velocity increased by more than 20%.

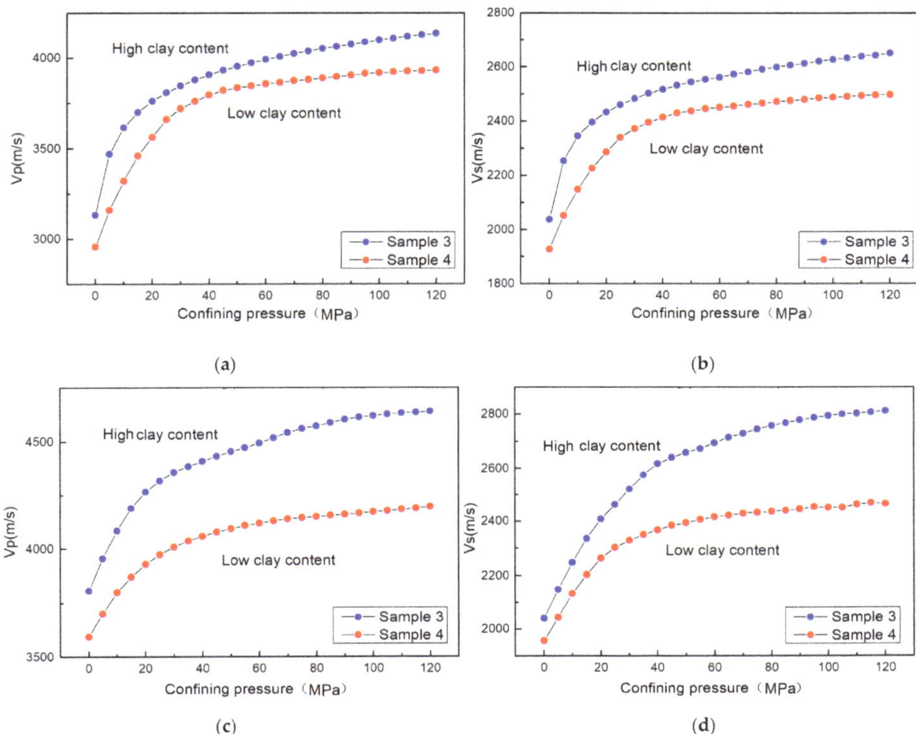

Figure 9. Relationship between P- and S-wave velocity and confining pressure, (**a**,**b**) are the relationship between wave velocity and pressure under dry conditions, and (**c**,**d**) are the relationship between wave velocity and pressure under completely saturated conditions.

We speculate that this was related to the elastic properties of the clay and its related pores. Under a low confining pressure, the velocities of both samples had a positive correlation with confining pressure due to the closure of plastic pores. When the pressure was greater than 40–60 MPa and the plastic pores were completely closed, the resistance of the rigid pores and skeleton to the pressure was found to affect the wave velocity. The resistance of rigid pores in mineral particles to pressure was also related to the hardness of minerals. The sample with low clay content had higher contents of stiffer quartz and feldspar minerals, and the resistance of rigid pores and skeleton to pressure was significantly higher than that of the high-clay sample. Therefore, the P- and S-wave velocity of the low-clay-content sample was significantly lower than that of samples with higher clay content under high pressure. It should be noted that the P- and S-wave velocity with the high-clay-content sample was higher than those with the low-content sample under 0 MPa confining pressure. This may be related to the mineral morphology. That is, the morphology of mineral and pore in rock have a great influence on the wave velocity [28]. In conclusion, the content of clay, quartz, and feldspar minerals in the rock has a secondary influence on the relationship between wave velocity and confining pressure. If the content of quartz and feldspar minerals are higher, the P- and S-wave velocity increases slowly with confining pressure under high conditions. On the contrary, if the clay content is higher, the P- and S-wave velocity increases greatly with confining pressure under certain high conditions. The phenomenon of false bright spot may be related to the change of clay content in a high-pressure formation.

The variable temperature experiment was carried out on two samples. The cross-plot of temperature and P- and S-wave velocity of two dry samples from 0 to 150 °C under 10 MPa confining pressure is shown in Figure 10. The P- and S-wave velocity of the two samples exhibited similar trends with temperature. However, the velocity reduction of the high-clay sample was greater than that of the low-clay sample (21% vs. 19%). Due to the expansion of clay itself and internal pores during the heating process, the expansion rate of clay increases with temperature [26,27]. As the range of softening of rock elastic properties increased, the decrease in the P- and S-wave velocity sample with high clay content became greater. Thus, for a high-temperature formation, the change of clay content may lead to impedance difference and a false bright spot.

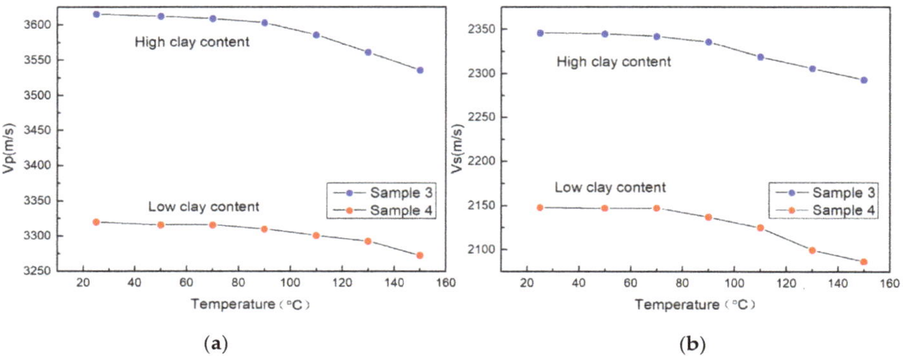

Figure 10. Cross-plot of temperature and (**a**) P- and (**b**) S-wave velocity.

5. Pore Pressure

We used the samples with large differences in porosity and minerals for analysis, and the experimental conditions were completely saturated by pressurized vacuum pumping. Figure 11 shows the relationship between P- and S-wave velocity and pore pressure of the two samples with large porosity differences under room temperature (confining pressure remained unchanged at 60 Mpa). Both samples had a negative correlation between wave velocity and pore pressure, but the P-wave velocity decrease of the high-porosity sample was greater than that of the low-porosity sample (8.5% vs. 6%).

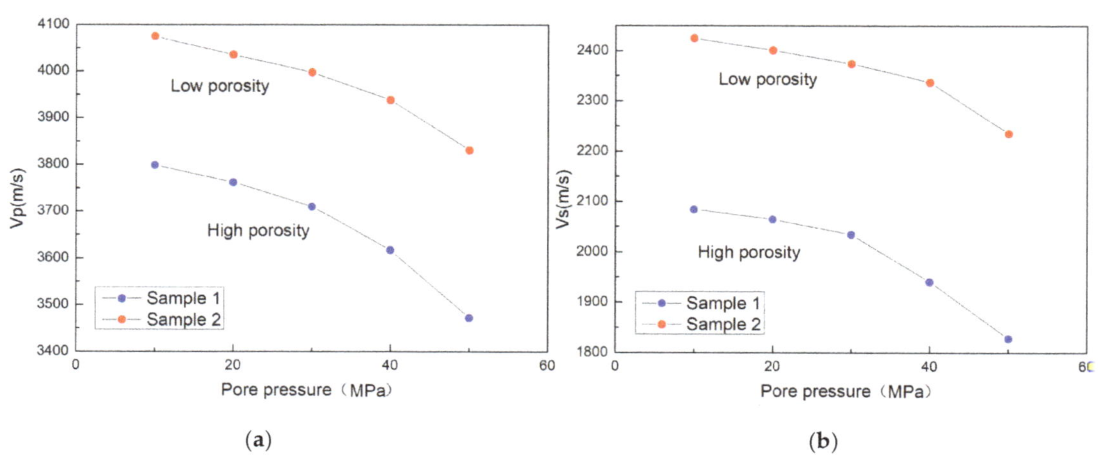

Figure 11. Cross-plot of pore pressure and (**a**) P- and (**b**) S-wave velocity.

Similarly, two samples with large differences in clay minerals and small differences in pores were analyzed. As shown in Figure 12, the P- and S-wave velocities of the samples decreased with the increase of pore pressure, but the decrease value was different and the Vs for the high clay content showed an even larger drop compared to that of Vp. The P-wave velocity of the high clay-content sample decreased by more than 7%, and that of the low-clay-content sample decreased by less than 4.5%. In the process of increasing pore pressure, due to the tight skeleton of the low-clay-content sample, the resistance of pores in the skeleton to pore pressure was higher than high-clay sample, which made the degree of pore expansion smaller and finally led to the reduction of their velocities in varying degrees. Combined with all the experimental results, we could draw a conclusion: Under the comprehensive influence of porosity and mineral composition, PT and P- and S-wave velocity differences appeared in the samples. The formation with high porosity, high clay content, high temperature, and overpressure was prone to an abnormal reduction of wave velocity, velocity and impedance difference with the compacted high-velocity mudstone layer or background mudstone layer, resulting in false bright spots. Similarly, the abnormal reduction of velocity of this formation may have also led to the reduction of impedance difference with gas-bearing formation, resulting in a dark point gas reservoir.

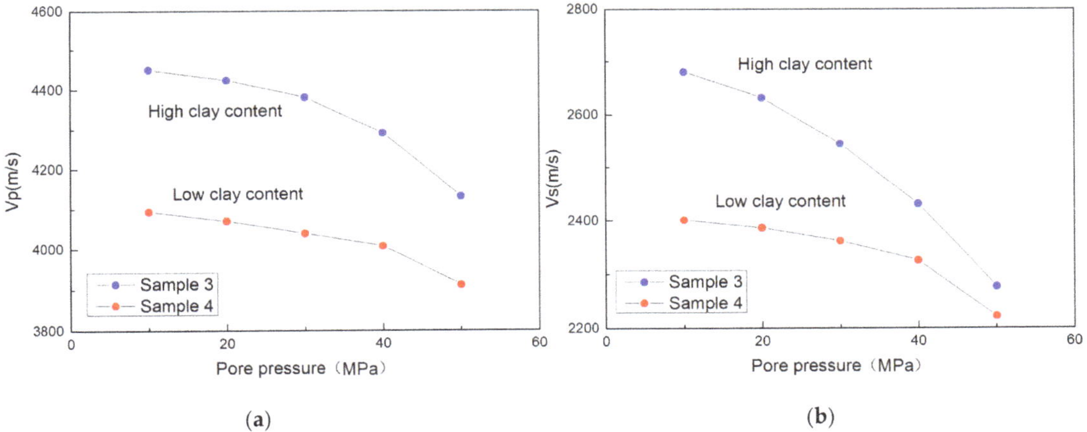

Figure 12. Cross-plot of pore pressure and (**a**) P- and (**b**) S-wave velocity.

6. Case Analysis and Theoretical Simulation

Figure 13 shows the DF1-X6 well-logging data of Meishan first segment team in Dongfang district. The drilling results showed that there was no effective reservoir in the target layer. The cable measure formation pressure and logging interpretation results showed that the mudstone layer at 3470 m had the characteristics of overpressure and high temperature, its pore pressure (4143.1 psi), temperature (141.33 °C), porosity (10.8%), and clay content (29.2%) were higher than those of the overlying silty mudstone (pore pressure 2077.34 psi, temperature 138.25, clay content 10.6%, porosity 6.7%), resulting in a low P-wave velocity characteristic. The impedance difference between the upper and lower layers appeared due to the same density and different velocity which led the false bright spot in the seismic profile. This phenomenon also confirmed our conclusion that high porosity, clay content, temperature, and overpressure formation could easily cause abnormal velocity reduction.

Figure 14 shows the LD23-X1 well-logging data of Yinggehai first segment team in Ledong district. The actual drilling results also revealed no effective reservoir. The cable and logging interpretation results showed that an abnormal overpressure point (pore pressure 6120.64 psi) in the high-porosity (20.7%) sandy mudstone (clay content 15.3%) appeared at

2316 m and presented a low P-wave velocity characteristic. The mudstone at the 2327 m overlying strata presented (porosity 11%) high clay content characteristics (33.1%). Due to the drainage and compaction of low-velocity mudstone, high-velocity mudstone was formed. The impedance difference between them caused the abnormal reflection of seismic response. This phenomenon also confirmed our view that high porosity and pore pressure samples easily reduced the P-wave velocity, and high-clay-content samples easily increased the wave velocity greatly under high pressure.

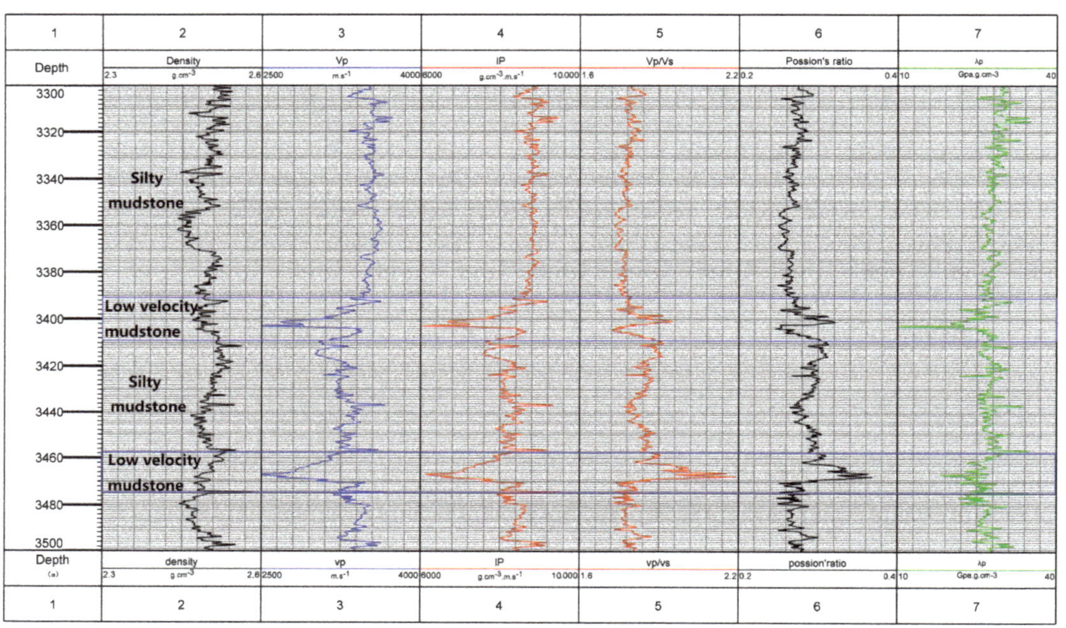

Figure 13. Logging curve of well DF1-X6.

Figure 15 shows the DF1-X12 well-logging data of Huangliu first segment team in Dongfang district. The cable formation pressure test results showed that 20 data points within the depth of 2633–2719 m had high temperature characteristics (greater than 113 °C). The mudstone interval at 2660–2673 m had higher clay content (19.6%) than other depth sections, resulting in a significant decrease in velocity under high-temperature conditions. The underlying sand layer (porosity 14.8%) showed higher porosity than the overlying sand layer (porosity 9.5%), and the velocity decreased slightly under high-temperature conditions. Low-velocity mudstone showed the characteristics of high density and low velocity, and sandstone showed the characteristics of high velocity and low density, resulting in the reduction of their impedance difference and the formation of a dark point gas reservoir. Generally, the mudstone near the reservoir cannot show low-velocity characteristics due to the reservoir compaction [4]. However, the results showed that dark point gas reservoirs may also occur between the reservoir and mudstone due to the influence of temperature, clay content, and porosity. This phenomenon was also confirmed by the experimental data.

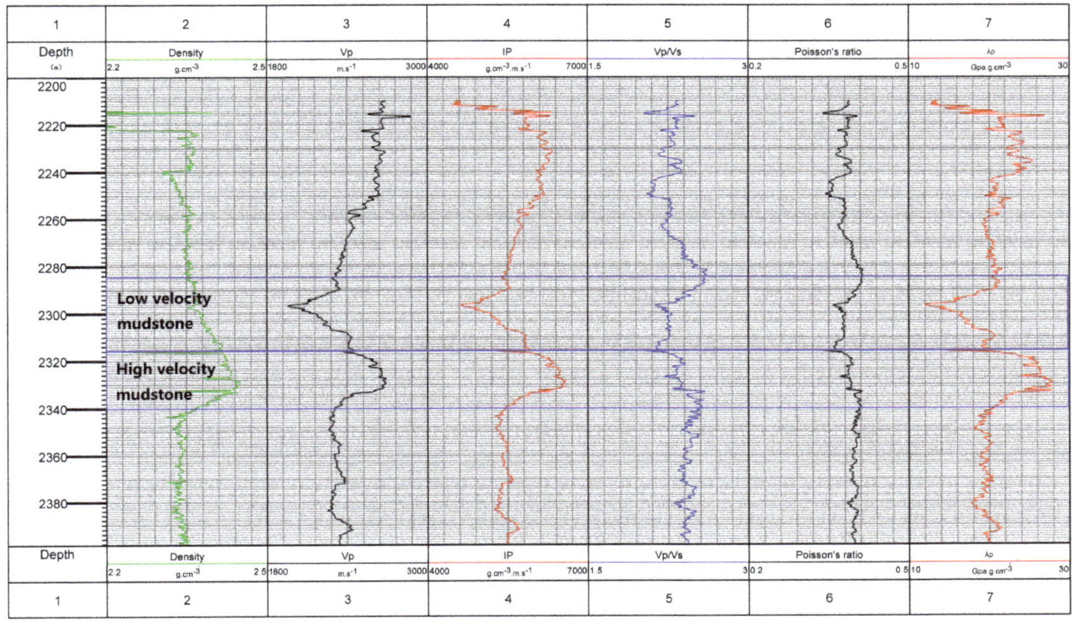

Figure 14. Logging curve of well LD23-X1.

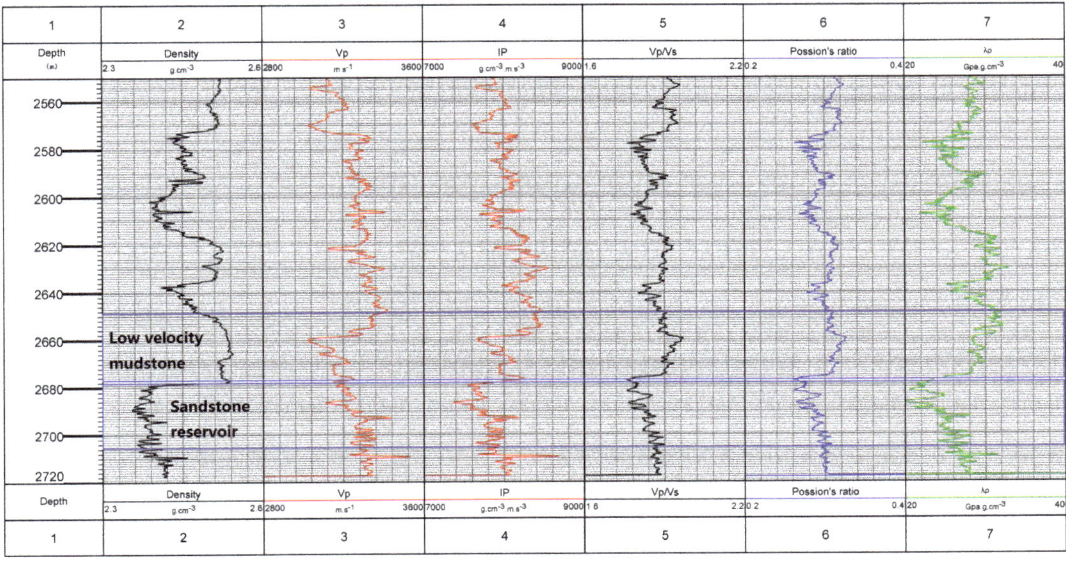

Figure 15. Logging curve of well DF1-X12.

We used the experimental results to predict the formation's P-wave velocity data through the fitting formula. For P-wave prediction of the logging data, the fitting relationship between velocity and PT was established according to the experimentally measured values. Moreover, in order to avoid the dispersion problem of saturated samples under high-frequency conditions [29–31], the relationship between effective stress and velocity was constructed, and the effective stress was equal to the difference between the confining pressure and pore pressure when the rock was completely saturated. The formula was as follows:

$$V = V_0 + A * \ln(p/p_0) + B * (T - T_0) \tag{1}$$

In the formula, V represents the fitting velocity, V_0 represents the initial velocity of the sample, p_0 and T_0 are the initial effective stress and initial temperature, respectively, and P and T are the actual stress and actual temperature, respectively.

The fitting results R^2 of the four samples were greater than 0.8, and the fitting coefficients are shown in Table 3.

Table 3. Fitting coefficients of four samples.

Sample Number	1	2	3	4
A (Effective pressure coefficient)	445.79	215.30	266.85	210.19
B (Temperature coefficient)	−0.89	−0.36	−0.46	−0.27

The coefficient reflects the differences in the sensitivity of pores and mineral components to PT. Compared with that of the low-porosity sample, the high-porosity sample had higher PT coefficients. The high coefficient reflected that the P- and S-wave velocity increased more with the confining pressure and decreased more with the increase of temperature and pore pressure. Compared with that of the low-clay-content sample, the high effective PT coefficient of the high-clay-content sample also reflected this feature. The fitting coefficient results also confirmed our experimental view that the velocity could vary significantly in high temperature and overpressure formation due to the influence of rock pores and minerals. Therefore, the traditional fitting formula could not accurately predict the P-wave data of the high confining pressure and temperature formation. We used a multiple linear regression method to establish the relationship between the PT coefficient and PC (porosity and clay content):

$$A = 20.74 \times \varphi + 0.05 \times V_{\text{clay}} - 73.36 \tag{2}$$

$$B = 0.35 - 0.048 \times \varphi - 0.003 \times V_{clay} \tag{3}$$

where A and B represent pressure and temperature coefficient, respectively, and ϕ and V_{clay} represent porosity and clay content, respectively.

Many studies have simulated carbonate or shale through the DEM model [32–35]. We also used this model to predict the P-wave velocity combined with cable formation pressure and debris XRD or logging interpretation data points. See Appendix A for the model formula. The main modeling processes were as follows: (1) Mixing clay and pores with water by the V-R-H model, (2) the remaining minerals were mixed by the V-R-H model, (3) the clay pores mixture as an inclusion was added into the mineral mixture by the DEM model, (4) the P-wave velocity was calculated according to the improved fitting formula. The clay and pore morphology were controlled by pore aspect ratio for velocity of initial pressure, which was obtained by exhaustive inversion. Figure 16 shows the P-wave velocity of the three wells, and the bluepoints are the prediction data. As shown in the figure, we have achieved reliable prediction results.

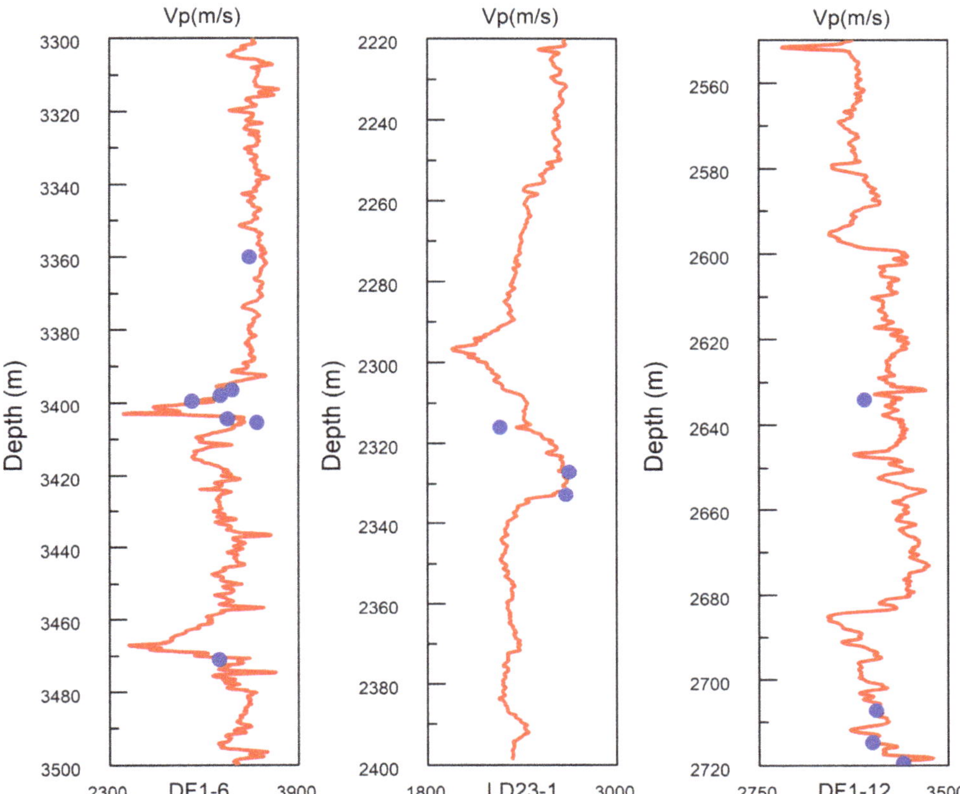

Figure 16. P-wave velocity prediction of the three wells (blue dots are the prediction data).

7. Conclusions

In this paper, the factors influencing P- and S-wave velocity differences under high PT conditions were analyzed from the perspectives of porosity, pore structure, and mineral composition. Combined with the experimental results, the causes of false "bright spot" and dark spot gas reservoirs in the seismic data in the study area were summarized. The main conclusions are as follows:

1. Porosity had important influence on the relationship between P- and S-wave velocity and PT. The P- and S-wave velocity of the high-porosity sample increased greatly with confining pressure under the condition of 0–60 MPa and increased slowly under 60–120 MPa. The P-wave velocity increased more than 70%. The velocity trend of the low-porosity sample under low pressure was similar to that of the high-porosity sample, but the velocity hardly increased under high pressure, and the P-wave velocity generally increased less than 50%. When the pore pressure increased to 60 MPa, the P- and S-wave velocity decrease of the high-porosity sample was significantly greater than that of the low-porosity sample. The P-wave velocity decreased by more than 8.5%, and by less than 6% for the low-porosity sample. Under high temperature (150 °C), the P-wave velocity of the high-porosity sample decreased by more than 7%, and that of the low-porosity sample decreased by 4%.

2. Mineral composition also had certain influence. Under the high confining pressure of 60–120 MPa, the P- and S-wave velocity of the low-clay-content sample gradually slowed down with the increase of pressure, and P-wave velocity increased by

16%, while the velocity of the high-clay-content sample still maintained a large increase range, and P-wave velocity increased by more than 20%. When pore pressure increased to 60 MPa, the P- and S-wave velocity of the high-clay-content sample decreased more than that of the low-clay-content sample, the P-wave velocity decreased more than 7%, and the P-wave velocity of the low-clay-content sample decreased less than 4.5%. Under high temperature (150 °C), the P- and S-wave velocity decrease of the high-clay-content sample was significantly higher than that of the low-clay-content sample. The P-wave velocity of the high-clay-content sample decreased by more than 21%, and that of the low-clay-content sample decreased by 14%.

3. The false bright spot and dark spot gas reservoirs of seismic data in Yinggehai basin were related to the differences of pores, clay content, pressure, and temperature of the upper and lower formations. High porosity and clay content formation easily caused abnormal P- and S-wave velocity reduction under high-temperature and overpressure conditions and easily caused a velocity rise under high confining pressure conditions. The velocity difference caused the change of impedance to form the phenomenon of a false bright spot or dark spot gas reservoir.

4. The coefficient of PT reflected the sensitivity of pores and skeleton to pressure and temperature. According to the change of PT coefficient, a modified P-wave velocity prediction method was proposed in this paper.

Author Contributions: Conceptualization, Z.L. and Z.Z.; methodology X.D.; software, X.L.; writing–original draft preparation, Z.L., writing–review and editing, X.D. All authors have read and agreed to the published version of the manuscript.

Funding: This research was funded by sponsorship of National Nature Science Foundation Project grant number [U19B2008].

Institutional Review Board Statement: Not applicable.

Informed Consent Statement: Not applicable.

Data Availability Statement: Not applicable.

Conflicts of Interest: The authors declare no conflict of interest.

Appendix A.

Appendix A.1. Voight/Reuss and the V-R-H Model

The Voigt and Reuss model [36] provides the upper and lower limit of the elastic modulus of the equivalent medium. The formulae are as follows:

$$K_v = \sum_{i=1}^{N} K_i * V_i \qquad K_R^{-1} = \sum_{i=1}^{N} K_i^{-1} * V_i$$
$$u_v = \sum_{i=1}^{N} u_i * V_i \qquad V_R^{-1} = \sum_{i=1}^{N} u_i^{-1} * V_i \tag{A1}$$

Hill [37] regarded the arithmetic average of the above two models as the equivalent elastic modulus of the medium:

$$K = \frac{K_V + K_R}{2}$$
$$\mu = \frac{\mu_V + \mu_R}{2} \tag{A2}$$

where K_i and u_i represent the bulk modulus and shear modulus of each component, V_i represents the volume content of each component, K_v and u_v represent the upper limit of bulk modulus and shear modulus of medium, K_R and V_R represent the lower limit of bulk modulus and shear modulus of medium, and K and u represent the bulk modulus and shear modulus of medium.

Appendix A.2. Hashin–Shtrikman Averaging

The upper and lower limits of equivalent modulus by defining stiffer pore shape and soft pore shape are given as [38]:

$$K^{HS+} = K_1 + \frac{f_2}{(K_2 - K_1)^{-1} + f_1 (K_1 + \frac{4}{3}\mu_1)^{-1}}$$

$$\mu^{HS+} = \mu_1 + \frac{f_2}{(\mu_2 - \mu_1)^{-1} + \frac{2f_1(K_1 + 2\mu_1)}{5\mu_1(K_1 + \frac{4}{3}\mu_1)}}$$

(A3)

where f_1 and f_2 represent the volume content of each component, K_1 and K_2 represent bulk modulus of each component, and μ_1 and μ_2 represent shear modulus of each component. The formula determines the upper and lower limits by determining the stiffer and soft of the two components.

Appendix A.3. Different Effective Medium

The different effective medium (DEM) model assumes that the background matrix is a solid phase, continuing to add the containing phase into the background matrix until the component content reaches the theoretical value. The formula is given by Berryman [39]:

$$(1 - y)\frac{d}{dy}[K^*(y)] = (K_2 - K^*)P^{(*2)}(y)$$

$$(1 - y)\frac{d}{dy}[\mu^*(y)] = (\mu_2 - \mu^*)Q^{(*2)}(y)$$

(A4)

where y represents volume content of inclusion, K^* represents bulk modulus of mixture, μ^* represents shear modulus of inclusion, K_2 represents bulk modulus of inclusion, μ_2 represents shear modulus of inclusion, and $P^{(*2)}$ and $Q^{(*2)}$ represent the influence factor of changing the shape of inclusion.

References

1. Pan, G.C.; Pei, J.; Zhou, J.; Wang, L.F.; Yu, J.F. An Analysis of Flat Spot Features on a Gas-Water Interface in the Middle-Deep Overpressure zone, Yinggehai Basin. *China Offshore Oil Gas* **2014**, *26*, 5. (In Chinese)
2. Pei, J.; Yu, J.; Wang, L.; Hao, D.F.; Liu, F. Key Challenges and Strategies for the Success of Natural Gas Exploration in Mid-Deep Strata of the Yinggehai Basin. *Acta Pet. Sin.* **2011**, *46*, 145–146. (In Chinese)
3. Zhou, J.X.; Liu, W.W.; Ma, G.K.; Wang, L.F.; Liu, B. Fine Description and Prediction of Seismic Attributes of HPHT Gas Reservoir in the Yinggehai Basin. *Nat. Gas Ind.* **2013**, *33*, 7–11. (In Chinese)
4. Pei, J.X.; Pan, G.C.; Zhu, P.Y.; Liu, F. Identification of Low Velocity Mudstone in the Middeep Overpressure zone, Yinggehai Basin. *Oil Geophys. Prospect.* **2016**, *51*, 10. (In Chinese)
5. Kern, H. The Effect of High Temperature and High Confining Pressure on Compressional Wave Velocities in Quartz-Bearing and Quartz-Free Igneous and Metamorphic Rocks. *Tectonophysics* **1978**, *44*, 185–203. [CrossRef]
6. Scheu, B.; Kern, H.; Spieler, O.; Dingwell, D. Temperature Dependence of Elastic P- and S-Wave Velocities in Porous Mt. Unzen Dacite. *J. Volcanol. Geotherm. Res.* **2006**, *153*, 136–147. [CrossRef]
7. Sone, L.T.; Liu, Z.H.; Liu, Z.H.; Wang, Q. Elastic Anisotropy Characteristics of Tight Sands under Different Confining Pressure and Fluid Saturation States. *Chin. J. Geophys.* **2015**, *58*, 3401–3411.
8. Motra, H.B.; Zertani, S. Influence of Loading and Heating Processes on Elastic and Geomechanical Properties of Eclogites and Granulites. *J. Rock Mech. Geotechnol. Eng.* **2018**, *10*, 127–137. [CrossRef]
9. Kern, H. P- and S-Wave Anisotropy and Shear-Wave Splitting at Pressure and Temperature in Possible Mantle Rocks and Their Relation to the Rock Fabric. *Phys. Earth Planet. Inter.* **1993**, *78*, 245–256. [CrossRef]
10. Eberhart-Phillips, D.; Han, D.-H.; Zoback, M.D. Empirical Relationships among Seismic Velocity, Effective Pressure, Porosity, and Clay Content in Sandstone. *Geophysics* **1989**, *54*, 82–89. [CrossRef]
11. Khaksar; Griffiths; McCann. Compressional- and Shear-Wave Velocities as a Function of Confining Stress in Dry Sandstones. *Geophys. Prospect.* **1999**, *47*, 487–508. [CrossRef]
12. Mavko, G.; Mukerji, T.; Dvorkin, J. *The Rock Physics Handbook: Tools for Seismic Analysis of Porous Media*; Cambridge University Press: Cambridge, UK, 2009.
13. Liu, Z.; Zhang, F.; Li, X. Elastic Anisotropy and Its Influencing Factors in Organic-Rich Marine Shale of Southern China. *Sci. China Earth Sci.* **2019**, *62*, 1805–1818. [CrossRef]
14. Deng, J.X.; Wang, H.; Zhou, H.; Liu, Z.H.; Song, L.T.; Wang, X.B. Microtexture, Seismic Rock Physical Properties and Modeling of Longmaxi Formation Shale. *Chin. J. Geophys.* **2015**, *58*, 2123–2136.

15. Han, D.; Nur, A.; Morgan, D. Effects of Porosity and Clay Content on Wave Velocities in Sandstones. *Geophysics* **1986**, *51*, 2093–2107. [CrossRef]
16. Salih, M.; Reijmer, J.; El Husseiny, A.; Bashri, M.; Eltom, H.; Mukainah, H.; Kaminski, M. Controlling Factors on Petrophysical and Acoustic Properties of Bioturbated Carbonates: (Upper Jurassic, Central Saudi Arabia). *Appl. Sci.* **2021**, *11*, 5019. [CrossRef]
17. Jaballah, J.; Reijmer, J.; El-Husseiny, A.; Le, J.; Hairabian, A.; Slootman, A. Physical Properties of Cretaceous to Eocene Platform-to-Basin Carbonates from Albania. *Mar. Pet. Geol.* **2021**, *128*, 105022.
18. Dvorkin, J.; Alabbad, A. Velocity–porosity–mineralogy Trends in Chalk and Consolidated Carbonate Rocks. *Geophys. J. Int.* **2019**, *219*, 662–671.
19. Dvorkin, J.; Walls, J.; Davalos, G. Velocity-Porosity-Mineralogy Model for Unconventional Shale and Its Applications to Digital Rock Physics. *Front. Earth Sci.* **2021**, *8*, 613716. [CrossRef]
20. Reijmer, J.J.G.; Blok, C.N.; El-Husseiny, A.; Kleipool, L.M.; Hogendorp, Y.C.K.; Alonso-Zarza, A.M. Petrophysics and Sediment Variability in a Mixed Alluvial to Lacustrine Carbonate System (Miocene, Madrid Basin, Central Spain). *Depos. Rec.* **2021**, *8*, 317–339. [CrossRef]
21. Assefa, S.; McCann, C.; Sothcott, J.; Astin, T.; Johnstad, S. The Effects of Porosity—Pore-Fluid and Mineralogy on VpNs in Carbonate Rocks. In Proceedings of the 61st EAGE Conference and Exhibition, Helsinki, Finland, 7–11 June 1999. [CrossRef]
22. El-Husseiny, A.; Vega, S.; Nizamuddin, S. The Effect of Pore Structure Complexity and Saturation History on the Variations of Acoustic Velocity as Function of Brine and Oil Saturation in Carbonates. *J. Pet. Sci. Eng.* **2019**, *179*, 180–191. [CrossRef]
23. Grana, D. Pressure–velocity Relations in Reservoir Rocks: Modified Macbeth's Equation. *J. Appl. Geophys.* **2016**, *132*, 234–241.
24. Neto, I.A.L.; Misságia, R.M.; Ceia, M.; Archilha, N.L.; Oliveira, L.C. Carbonate Pore System Evaluation Using the velocity–porosity–pressure Relationship, Digital Image Analysis, and Differential Effective Medium Theory. *J. Appl. Geophys.* **2014**, *110*, 23–33. [CrossRef]
25. Wu, Z.; Wang, W. First-Principles Calculations of Elasticity of Minerals at High Temperature and Pressure. *Sci. China Earth Sci.* **2016**, *59*, 1107–1137. [CrossRef]
26. Zhang, Y.L.; Sun, Q.; Li, J.X.; Zhang, W.Q. Pore and Mechanical Characteristics of High-Temperature Bakeed Clay. *Chin. J. Rock Mech. Eng.* **2015**, *34*, 1480–1488. (In Chinese)
27. Mao, H.; Qiu, Z.S.; Huang, W.A.; Shen, Z.H.; Yang, L.Y.; Zhong, H.Y. The Effects of Temperature and Pressure on the Hydration Swelling Characteristics of Clay Mineral. *Pet. Drill. Technol.* **2013**, *41*, 56–61. (In Chinese)
28. Bandyopadhyay, K. Seismic Anisotropy-Geological Causes and Its Implications to Reservoir Geophysics. Ph.D. Thesis, Stanford University, Stanford, CA, USA, 2009.
29. Cheng, W.; Ba, J.; Fu, L.-Y.; Lebedev, M. Wave-Velocity Dispersion and Rock Microstructure. *J. Pet. Sci. Eng.* **2019**, *183*, 106466. [CrossRef]
30. Ba, J.; Xu, W.; Fu, L.-Y.; Carcione, J.M.; Zhang, L. Rock Anelasticity Due to Patchy Saturation and Fabric Heterogeneity: A Double Double-Porosity Model of Wave Propagation. *J. Geophys. Res. Solid Earth* **2017**, *122*, 1949–1976. [CrossRef]
31. Zhang, L.; Ba, J.; Carcione, J.M. Wave Propagation in Infinituple-Porosity Media. *J. Geophys. Res. Solid Earth* **2021**, *126*, 2020021266. [CrossRef]
32. Salih, M.; Reijmer, J.J.; El-Husseiny, A. Diagenetic Controls on the Elastic Velocity of the Early Triassic Upper Khartam Member (Khuff Formation, Central Saudi Arabia). *Mar. Pet. Geol.* **2020**, *124*, 104823. [CrossRef]
33. Francois, F.; Matthieu, P.; Quentin, V.; Thomas, T.; Fei, H.; Emmanuelle, P.; Jean, B.; Philippe, L.; Hairabian, A. The Equivalent Pore Aspect Ratio as a Tool for Pore Type Prediction in Carbonate Reservoirs. *AAPG Bull.* **2018**, *102*, 1343–1377.
34. Jafarian, E.; Kleipool, L.; Scheibner, C.; Blomeier, D.; Reijmer, J. Variations in Petrophysical Properties of Upper Palaeozoic Mixed Carbonate and Non-Carbonate Deposits, Spitsbergen, Svalbard Archipelago. *J. Pet. Geol.* **2016**, *40*, 59–83. [CrossRef]
35. Qian, K.; Zhang, F.; Chen, S.; Li, X.; Zhang, H. A Rock Physics Model for Analysis of Anisotropic Parameters in a Shale Reservoir in Southwest China. *J. Geophys. Eng.* **2016**, *13*, 19–34. [CrossRef]
36. Reuss, A. Berechung der Fliessgrenzen von Mischkristallen auf Grund der Plastizitatsbedingung fur Einkristalle. *Z. Angew. Math. Phys.* **1929**, *9*, 49–58.
37. Hill, R. The Elastic Behavior of Crystalline Aggregate. *Proc. Phys. Soc.* **1952**, *65*, 349–354.
38. Hashin, Z.; Strikman, S. A Variation Approach to the Theory of Effective Magnetic Permeability of Multiphase Materials. *J. Appl. Phys.* **1962**, *33*, 3125–3131.
39. Berryman, J.G. Single-scattering Approximations for Coefficients in Biot's Equations of Poroelasticity. *J. Acoust. Soc. Am.* **1992**, *91*, 551–571. [CrossRef]

Article

Stratigraphic Constraints on Sandy Conglomerates in Huanghekou Sag, Bohai Bay Basin, via In Situ U-Pb Dating of Vein Calcite and Detrital Zircons, and XRD Analysis

Wei Wang [1,2,*], Xianghua Yang [1,2,*], Hongtao Zhu [1,2] and Li Huang [1,2]

[1] School of Earth Resources, China University of Geosciences, Wuhan 430074, China; htzhu@cug.edu.cn (H.Z.); 1202020021@cug.edu.cn (L.H.)

[2] Key Laboratory of Tectonics and Petroleum Resources, Ministry of Education, China University of Geosciences, Wuhan 430074, China

* Correspondence: wang.wei@cug.edu.cn (W.W.); xhyang@cug.edu.cn (X.Y.)

Abstract: The discovery of the BZ19-6 large-scale condensate gas field illustrates the great potential of the sandy conglomerate reservoirs in the Bohai Bay Basin. However, the stratigraphic correlation of the sandy conglomerate sequence in northern Huanghekou Sag remains a challenge due to the lack of syn-depositional volcanic layers and biostratigraphic constraints. The challenge limits understanding the regional strata distribution and further exploration deployment. In this study, we conducted in situ U-Pb dating of vein calcite and detrital zircons of the sandy conglomerate samples from borehole BZ26-A. The vein calcite age and the youngest age of detrital zircons provide the upper and lower bounds of the depositional age, respectively. We also correlated the samples with those from well-understood strata through a comparison of XRD mineral components. The absolute age of 47.0 Ma of the vein calcite and the youngest detrital zircon age of 103.5 Ma suggest the sedimentary sequence is supposed to be referred to as the Kongdian Formation (65–50.5 Ma). The XRD data and petrological analysis suggest that the lithostratigraphy of the Kongdian Formation in Huanghekou Sag could be divided into at least three members, with Member 3 consisting of red sediment deposited in a hot and dry climate; Member 2 and Member 1 deposited as fan delta with major parent rock of Mesozoic volcanic rocks and Precambrian meta-granitoid, respectively. Member 1 shows significant potential for energy exploration due to high brittle mineral components and fracture development.

Keywords: sandy conglomerate; Huanghekou Sag; stratigraphic correlation; calcite U-Pb dating; detrital zircon

Citation: Wang, W.; Yang, X.; Zhu, H.; Huang, L. Stratigraphic Constraints on Sandy Conglomerates in Huanghekou Sag, Bohai Bay Basin, via In Situ U-Pb Dating of Vein Calcite and Detrital Zircons, and XRD Analysis. *Energies* **2022**, *15*, 3880. https://doi.org/10.3390/en15113880

Academic Editors: Gan Feng, Qingxiang Meng, Gan Li and Fei Wu

Received: 23 March 2022
Accepted: 19 May 2022
Published: 24 May 2022

Publisher's Note: MDPI stays neutral with regard to jurisdictional claims in published maps and institutional affiliations.

1. Introduction

The potential of the sandy conglomerate sequence as reservoirs has been proved by the discovery of the BZ19-6 large-scale condensate gas field in Bozhong Sag, Bohai Bay Basin, China, with reserves exceeding 100 billion cubic meters. The success lights the progress in studies of the sandy conglomerate in the Paleocene Kongdian Formation [1,2]. This is the first time large-scale thick fractured-porous sandy conglomerate reservoirs in the Kongdian Formation in Bohai Bay Basin have been discovered, which breaks through the previous forbidden area of deep-buried sandy conglomerates and greatly expands the exploration field of Paleogene oil and gas [1]. However, understanding the stratigraphic distribution of the sandy conglomerate sequence remains a challenge due to the interpretation ambiguity of deep seismic data (>3500 m), paucity of biostratigraphic information, and lack of reference laminae, such as mudstone or volcanic layers [3,4]. These characteristics limit the exploration process, especially in expanding the area from Bozhong Sag to the Huanghekou Sag in the south (Figure 1). Thus, new methods of stratigraphic correlation and further understanding of the Kongdian Formation are needed for future exploration.

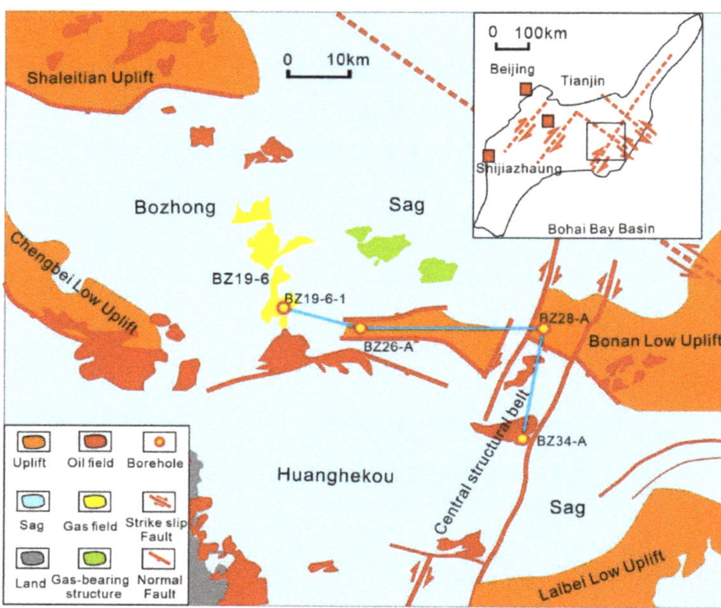

Figure 1. Geological map of the study area showing major tectonic elements, oil-gas bearing structures, and boreholes.

Although U-Pb dating of carbonate minerals, such as calcite and dolomite, is technically challenging due to low U concentrations (<10 ppm), the technique has been developing fast and is widely used in studies on carbonate diagenesis and brittle faulting in recent years [5–7]. However, the powerful method, which could provide the absolute time of precipitation of carbonate minerals, is seldom used in studies on clastic rocks. On the contrary, detrital zircon geochronology is widely used to constrain the maximum depositional age of a clastic unit, where the youngest age component provides the earliest possible age of deposition [8–10]. Since carbonate, especially calcite veins and detrital zircons, are widespread in sedimentary rocks, in this study, we try to conduct stratigraphic correlation through the in situ U-Pb dating of both vein calcite and detrital zircons, which provide the upper and lower bounds of the depositional age of the host rock, respectively. In addition, we also conducted XRD major and clay mineral analysis on the inter-gravel matrix of the sandy conglomerate samples. The mineral components containing information on parent rocks, diagenetic process, and paleoclimate were used for further correlation. The results could strengthen the understanding of the Kongdian Formation in the Bohai Bay Basin and shed light on similar correlating work in other areas.

2. Geological Setting

The Bohai Bay Basin is a significant hydrocarbon-bearing offshore basin located in Eastern China, which developed on the North China Block under the control of both Cenozoic rifting and strike-slip tectonics, along the Tan-Lu Fault Zone (Figure 1) [11,12]. The Paleocene of the basin can be divided into Kongdian, Shahejie and Dongying Formations (Figure 2), with stratigraphic data from the drilled values and bottom ages from the Editorial Committee of Petroleum Geology of China (1987) [13]. The Kongdian Formation ($E_{1-2}k$, bottom age of 65 Ma) and the fourth member of the Shahejie Formation (E_2s_4, bottom age of 50.5 Ma) were deposited in the initial extensional stage from the Paleocene to early Eocene. The deposition is characterized by fan delta sandy conglomerates and shallow lacustrine gray mudstones, accompanied by alluvial fan and lacustrine red conglomerates and mudstones [2,14]. The third member of the Shahejie Formation (E_2s_3), with a bottom age of 42 Ma, deposited during the early Eocene, is characterized by middle to deep lacustrine facies, with local fan delta deposits [13,15] Member two (bottom age of 38 Ma) to member one (bottom age of 36 Ma) of the Shahejie Formation ($E_2s_2 \sim E_2s_1$) are deposited under a relatively stable tectonic setting with deposition of lacustrine mudstones and fan delta pebbly sandstones. The Dongying Formation (E_3d), deposited during the Oligocene, was characterized by shallow lacustrine fine sandstones, interbedded with gray mudstones and delta plain pebbly sandstones interbedded with mudstone [13,15].

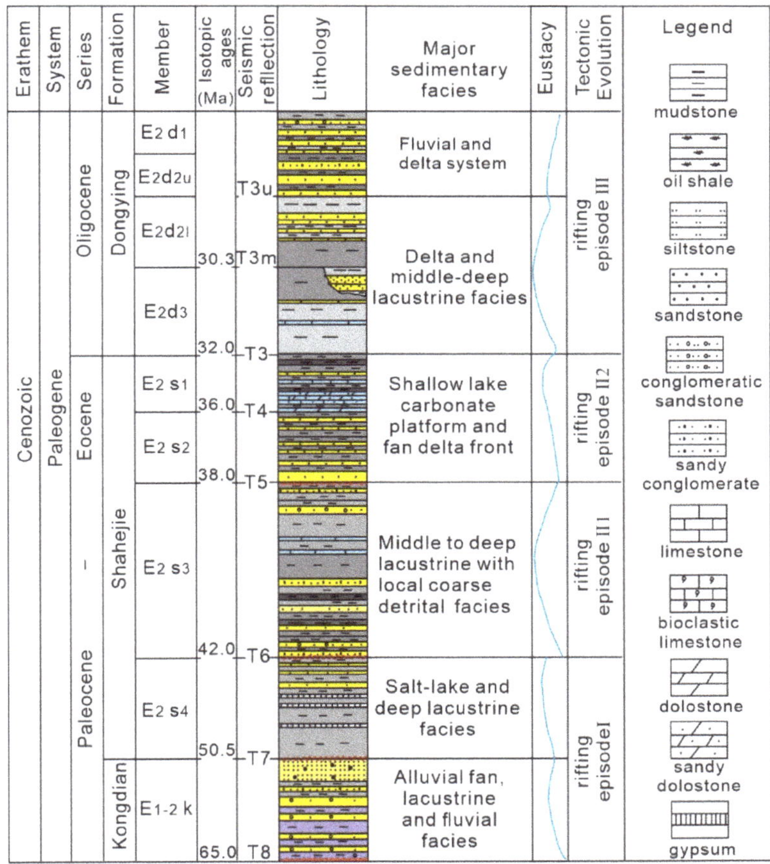

Figure 2. Generalized Cenozoic stratigraphy, facies, and tectonic events and stages of the Bohai Bay Basin.

The BZ19-6 Structure of Bozhong Sag, located between the Chengbei Low Uplift and the Bonan Low Uplift in the southwestern Bohai Sea (Figure 1), is an anticline where the Kongdian Formation is widely distributed. The major reservoirs of the Kongdian Formation are sandy conglomerates deposited in fan delta facies [1], accounting for an important part of the super-huge condensate oil field, with reserves of over 100 billion m^3 [1,2]. To the south of the BZ19-6 structure, the southern steep slope zone of the Bonan Low Uplift is a successive area for the exploration of deep-buried strata from Bozhong Sag to Huanghekou Sag, but with an ambiguous understanding of the stratigraphic distribution. BZ26-A is the only structure with boreholes and cores in this area (Figure 1), which uncovers a thick sandy conglomerate sequence with uncertain depositional ages due to a lack of biostratigraphic data and syn-depositional volcanic layers [16]. The absence of distinguishable sporopollen and reference lamina limits the understanding of the borehole and further strata distribution.

Fortunately, three intervals in the sandy conglomerates were cored from the borehole BZ26-A (Figure 3). The sedimentary and petrologic analysis suggests the sediment was deposited as a fan delta facies. In these intervals, abundant tectonic fissures developed and were partly filled with secondary minerals [16]. The C2 core interval shows the development of calcite veins filling parts of the tectonic fissures (Figure 4A). In the east, boreholes BZ28-A and BZ34-A were drilled into the Kongdian Formation on the Bonan Low Uplift and Central Structural Belt [14,15], respectively, due to relatively shallow buried depths on structural highs (Figures 1 and 3). Two intervals were cored from the conglomerate interval (~400 m in thickness) of BZ28-A. The core D1 consists of yellow-grey sandy conglomerates, while D2 shows a red to brown color (Figure 3). One interval was cored from borehole BZ34-A, which shows red to brown mixed mud and gravels (Figure 3). Both the boreholes revealed red to brown sediments (D2 and E1), which are considered reference lamina of the Kongdian Formation in the Bohai Bay Basin [14,17].

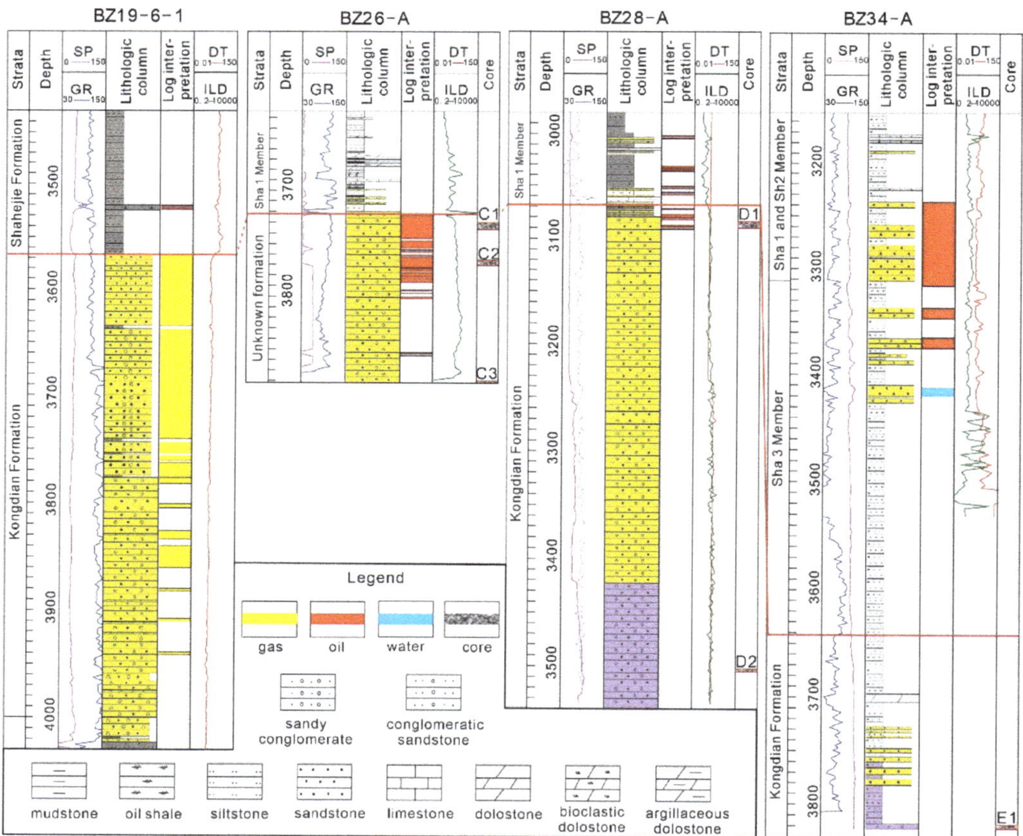

Figure 3. Stratigraphic section showing the borehole lithologies and the positions of sequence boundaries and core intervals.

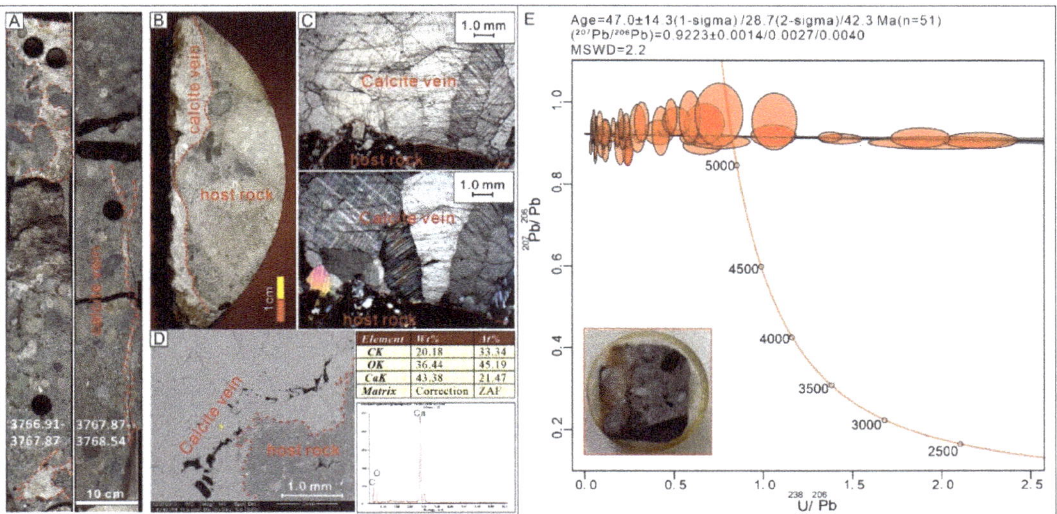

Figure 4. (**A**) Photos of part of the C2 core, BZ26-A with vein calcite marked; (**B**) photo of core showing contact relation of calcite vein and wall rock; (**C**) calcite vein micrographs under polarized (upper) and orthogonal light(lower); (**D**) back-scattering image of calcite vein and wall rock with residual pore and energy spectrum (EDS); (**E**) the chosen 2.5 cm-diameter mount of calcite vein and Tera-Wasseerburg ^{238}U/^{206}Pb-^{207}Pb/^{206}Pb Concordia plot for the vein calcite sample.

3. Sample and Method

3.1. In Situ U-Pb Dating of Vein Calcite

We collected four samples of calcite veins developed in the host rock of the sandy conglomerate sequence from the core C2 of borehole BZ26-A (Figure 3). Observation of cores, thin section analysis, and backscattered electron images with EDS show relatively homogeneous calcite filling the tectonic fissures of the host rock (Figure 4A–D).

LA-ICP-MS elemental pre-screening and in-situ U-Pb dating were conducted at the Radiogenic Isotope Facility, the University of Queensland, Australia. The four samples went through a lapidary process and were made into 2.5 cm diameter mounts (Figure 4E) and then assembled into a Laurin Technic sample holder for pre-screening for ^{238}U, ^{232}Th, ^{208}Pb, ^{207}Pb, ^{206}Pb ion beams and other key major and trace elements. In-situ element analysis was conducted using an ASI RESOlution SE 193 nm ultraviolet (UV) ArF-Excimer laser ablation system coupled with a Thermo I Cap-RQ quadrupole ICPMS. Laser repetition rate and energy levels were set to 15 Hz and 3 J/cm^2, respectively. The diameter of the laser spot was 100 µm. NIST-612 glass standard was employed for normalization and ^{43}Ca was used as an internal standard. The data were processed via Iolite software [18]. Areas with a range of ^{238}U/^{206}Pb correlated with ^{207}Pb/^{206}Pb, possibly defining a reference lower-intercept age on the Tera-Wassenburg Concordia diagram, were chosen as suitable for following U-Pb dating. Areas with low U/Pb ratios or extremely low U were excluded.

Further, 200 µm diameter spots were selected around the locations where high U/Pb ratios had been detected for the subsequent U/Pb dating. NIST-614 glass standard was used to correct for ^{207}Pb/^{206}Pb fractionation and instrument drift in the ^{238}U/^{206}Pb ratio. Typical sensitivity achieved was ~180,000 cps/ppm ^{238}U.

A matrix-matched calcite standard, AHX-1D (238.2 ± 1.5 Ma U-Pb age), was used to correct for matrix-related bias in ^{238}U/^{206}Pb ratios (unpublished). Another calcite standard of PDTK-2 (153.7 ± 1.7 Ma by isotope dilution, unpublished) was used as secondary reference material to ensure accuracy. Raw data were processed to obtain a range of

^{238}U/^{206}Pb and ^{207}Pb/^{206}Pb ratios, which may define a lower-intercept age on the Tera-Wassenburg Concordia diagram using the online version of IsoplotR of Vermeesch [19]. Uncertainties of calculated ages were reported at the 2σ level. During the period when our samples were measured, AHX-1D and PTKD-2 standards gave lower-intercept ages of 255.5 ± 2.2 Ma and 158.0 ± 2.1 Ma, respectively.

3.2. U-Pb Dating of Detrital Zircon

We collected one sample from the C2 core interval (3770.36–3770.56 m) of borehole BZ26-A and separated it using standard mechanical, heavy liquids, and electromagnetic mineral separation techniques. Zircon grains were picked under a binocular microscope, mounted in epoxy, and polished for U-Pb spot analyses. Cathodoluminescence (CL) images were used to evaluate the internal structure of grains, so that grains with inclusions and/or fractures were excluded from our analysis. U-Pb dating of zircons was conducted by LA-ICP-MS at the Micro-Macro Geochemistry Technology (Langfang) Co., Ltd., Langfang, China. Laser sampling was performed using a NewWave UP-213 Nd: YAG solid-state laser. An Agilent 7900 ICP-MS instrument was used to acquire ion-signal intensities.

During the test, helium is used as carrier gas and argon as compensation gas to adjust the sensitivity. They are mixed and connected to ICP-MS through a T-connector. The spot size of the laser was set to 30 μm. The energy density and frequency were set to 10 J/cm^2 and 10 Hz. NIST SRM610 was used to yield the highest sensitivity, lowest oxidation, background, and stable signal to achieve optimal condition. Standard zircon 91,500 was used as the external standard for isotope ratio correction [20], and standard zircon Plesovice (337 Ma) was used as the monitoring blind sample [21]. We Use GLITTER 4.0 software to calculate isotope ratio and element content.

3.3. X-Raydiffraction (XRD) Analysis of Bulk Rocks and Clay Minerals

Nine samples from C1, C2, and C3 intervals of BZ26-A were collected for XRD major mineral analysis with six of them for further clay mineral analysis. Five samples from the D1 and D2 core intervals of BZ28-A, and two samples from the E1 core interval of BZ34-A, were collected for XRD major and clay mineral analysis for comparison. The details of sampling sites and depths are available in Table S3 in the Supplementary Materials.

Samples were crushed into pieces and gravels were artificially excluded. Rock pieces were ground with an agate mortar until the total particle size was less than 300 mesh (48 microns). The mineralogy of the bulk rock and the clay minerals prepared by the side-loading method was carried out by X-ray diffraction (XRD) using an X-ray diffractometer (SmartLab SE). The mineral compositions and their relative proportions of the bulk rocks and clay minerals in the purified clay samples were obtained using the Clayquan (2018 version) program. The XRD analysis covers minerals of quartz, K-feldspar, plagioclase, calcite, dolomite, siderite, pyrite, hematite, laumontite, anhydrite augite, ankerite, and total clay minerals. The clay minerals were further analyzed for division of illite, kaolinite, chlorite, illite/smectite mixed layer, and chlorite/smectite mixed layer.

4. Results

4.1. InSitu U-Pb Dating of Vein Calcite

In-situ element analysis of 165 spots with a diameter of 100 μm was conducted for pre-screening. The results show a ubiquitously low concentration of ^{238}U, less than 0.01 ppm. Four spots from sample 1 with relatively higher ^{238}U of 0.02–0.27 ppm were chosen for further in situ dating. The result of 51 valid 66 U-Pb dating spots shows a narrow range of ^{238}U/^{206}Pb that is correlated with ^{207}Pb/^{206}Pb, defining a reference lower-intercept age on the Tera-Wassenburg Concordia diagram of 47 ± 14.3 Ma (1 sigma), with relatively large uncertainty (Figure 4E). The complete dataset is available in Table S1 in the Supplementary Materials.

4.2. U-Pb Dating of Detrital Zircon

Of the 240 detrital zircon grains analyzed from the core sample, a total of 235 grains yielded U-Pb ages within the concordant criteria (>80%). In general, the sample contains major groups of Late Archean to Early Proterozoic (~2500 Ma), Late Early Proterozoic (~1900–1800 Ma), and Late Mesozoic (150–100 Ma) and a minor group of Late Paleozoic to Early Mesozoic (~200–350 Ma). The youngest zircon with an age of 103.5 ± 2.5 Ma shows typical oscillatory zoning (Figure 5). The complete dataset is available in Table S2 in the Supplementary Materials.

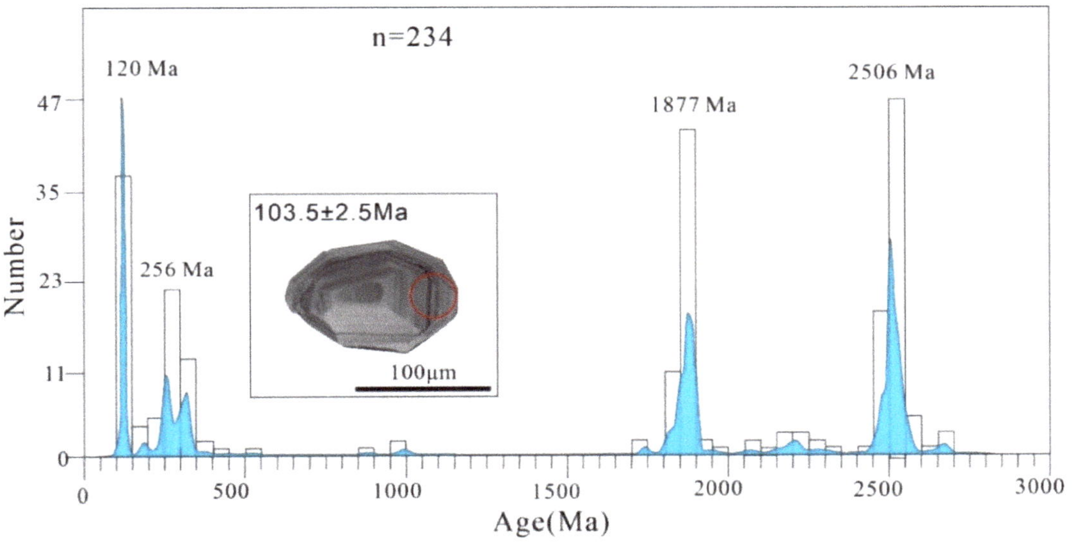

Figure 5. Kernel density estimation plot [22] of detrital zircon U-Pb ages of the sample from C2 core of borehole BZ26-A and CL image of the youngest zircon grain where the red circle marks the site of LA-ICPMS analysis. Data are discretized at a fixed interval (50 Ma).

4.3. XRD Analysis

The inter-gravel matrix of the sandy conglomerate is mainly composed of quartz, K-feldspar, plagioclase, augite, hematite, and clay minerals, which were converted into 100% and plotted (Figure 6). Six samples from cores C1 and C2 show similar components with dominant quartz (>60%), almost absent plagioclase, and approximately similar amounts of K-feldspar, augite, and clay minerals (less than 10%), respectively. Mineralogy of C3 and D1 is characterized by the occurrence of plagioclase, accounting for over ~20% and samples from red to brown cores of D2 and E1 show 4–6% hematite. The complete dataset is available in Table S3 in the Supplementary Materials.

The clay minerals are composed mainly of I/S (illite/smectite mixed layer), illite, kaolinite, and chlorite. Clay minerals of samples from cores C1 and C2 are mainly composed of I/S and illite. Samples from C3, D1, and D2 cores show high amounts of chlorite and kaolinite, while E1 is characterized by dominant I/S and illite (Figure 6).

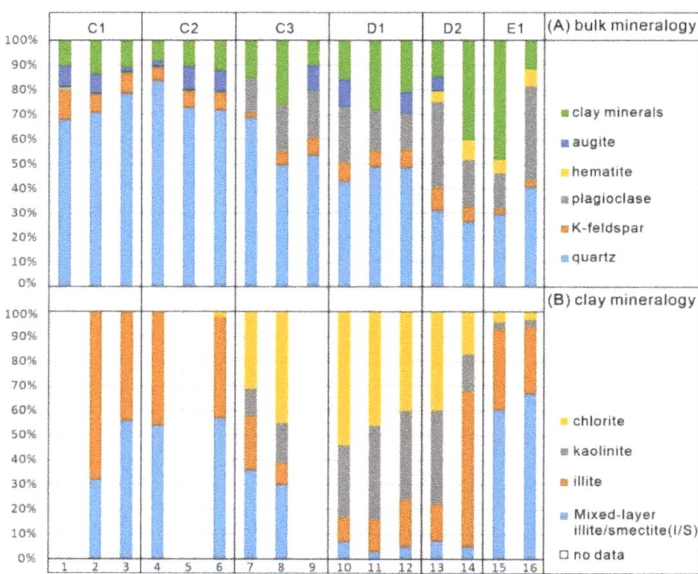

Figure 6. XRD major and clay mineral components.

5. Discussion

5.1. Depositional Age of the Thick Sandy Conglomerates from Borehole BZ26-A

Due to the lack of biostratigraphic information and syn-depositional volcanic layers, the depositional age of the sandy conglomerate could be achieved through constraints of absolute ages of minerals and correlation with well-understood samples. On one hand, the formation of tectonic fissures and following diagenesis took place after deposition, which suggests the time of secondary mineral precipitation could offer an upper bound for the depositional age. On the other hand, the detrital zircons in the sandy conglomerate were derived from the source areas, which crystallized before the deposition of the strata. Therefore, the youngest detrital zircon age could constrain the maximum depositional age [8,9].

The in situ U-Pb dating yields a major vein calcite age of 47.0 Ma, which falls in the time interval of 50.5–42.0 Ma, corresponding to the fourth member of the Shahejie Formation (Es$_4$) [13]. It indicates that the calcite precipitated synchronously with the deposition of Es$_4$ and the host rock of thick sandy conglomerates is supposed to be older than Es$_4$ and more likely to be part of the Kongdian Formation. However, it is worth noting that the narrow range of ^{238}U/^{206}Pb yields relatively high uncertainty, which makes the constraint less confident and it also indicates that the in situ U-Pb dating of carbonate minerals in clastic rocks remains a technical challenge.

The detrital zircon ages of the sandy conglomerates correspond well to the base rocks in the Bohai Bay Basin exposed on adjacent uplifts. It is composed of Precambrian grains with peaks at 2506 Ma and 1877 Ma, respectively, corresponding to the dominated age clusters in the North China Block [23–26], and the Mesozoic grains with a major peak at 120 Ma, corresponding to the Yanshanian tectonic event [27,28] (Deng et al., 2007; Wang et al., 2018) (Figure 5). The youngest age, 103.5 Ma, suggests no Cenozoic zircon grains were detected in the detrital zircons, and the strata are supposed to be deposited after 103.5 Ma, which provides a wide lower-bound constraint. Therefore, the depositional age of the sandy conglomerate sequence falls into the time interval of 103.5 Ma to 47.0 Ma. Combined with the Cenozoic geological setting, the sandy conglomerate sequence is more likely to be part of the Kongdian Formation, similar to those deposited in the BZ19-6 structure.

In addition, in the east of Bonan Low Uplift and Central Structural Belt, two boreholes with cores drilled into the Kongdian Formation and the distinguished red to brown deposits offer relatively continuous depositional records to compare. The mineral components in the matrix of the conglomerates are an integrated reflection of the source system in a certain period, which offers another proxy for the stratigraphic correlation.

5.2. Mineralogy of Kongdian Formation in Huanghekou Sag

As revealed by previous studies and the detrital zircon analysis in this study, the pre-Cenozoic basement consisting of Archaean–Proterozoic meta-granites, Early Paleozoic carbonate, and Mesozoic volcanic rocks from bottom to top, which were exposed on the Bonan Low Uplift, provided the sedimentary materials that were deposited in northern Huanghekou Sag [29]. The erosional process in chronological order under variable climate is supposed to be recorded by the sediments and could be traced by the change in major minerals of quartz, K-feldspar, plagioclase, and augite derived from the source area, and hematite, an indicator of climate.

The mineral components in the inter-gravel matrix from the sandy conglomerates are integrated reflections of the provenance system in a certain period. As shown in Figure 6, the XRD data could be well organized and divided into several groups. The red to brown strata of cores D1 from BZ28-A and E1 from BZ34-A are considered as the mark of the Kongdian Formation, where hematite (4–8%) is the cause of the color, reflecting a hot and dry climate [14]. Samples from core D1 (BZ28-A) and C3 (BZ26-A) show no hematite but approximately 20% of plagioclase, which is absent in the samples from C1 and C2 intervals of BZ26-A. The high contents of quartz and low contents of plagioclase from C1 and C2 suggest enhanced sources from the Precambrian meta-granites compared with the Mesozoic volcanic rocks with high plagioclase phenocrysts. The clay mineral analysis suggests a similar trend, where samples from D1, D2, and C3 show higher contents of chlorite than C1 and C2, which might be the result of the dissolution of Mesozoic volcanic rocks.

5.3. Implication on the lithostratigraphy of Kongdian Formation in Huanghekou Sag

The in situ U-Pb dating of both vein calcite and detrital zircons provides the upper and lower bounds of the depositional age of the sandy conglomerates, accumulated in the northern steep slope zone of Huanghekou Sag. Combined with the XRD mineral analysis, the lithostratigraphic units of the Kongdian Formation in Huanghekou Sag could be subdivided into at least three members (Table 1).

Table 1. Schematic correlation chart of the Kongdian Formation in Huanghekou Sag.

Kongdian Fm.	Geochronologic Constraints	Diagnostic Characteristics	Unknown BZ26-A	Known BZ28-A	Known BZ34-A
Member 1	(1) Vein Calcite age = 47 ± 14.3 Ma (2) Youngest detrital zircon age = 103.5 ± 2.5 Ma	(1) High quartz and K-feldspar with low plagioclase (2) Major provenance of Precambrian meta-granites	C1 and C2		
Member 2		(1) High plagioclase and high clay minerals (2) Major provenance of Mesozoic volcanic rocks	C3	D1	
Member 3		(1) Red to brown strata (2) Hot and dry climate		D2	E1

Member 3 of the Kongdian Formation was deposited in a hot and dry climate when the Huanghekou Sag was characterized by separated uplifts and depressions, with deposition of alluvial fans of red to brown sandy conglomerates. With the expansion of the lacustrine basin and enhanced faulting activity, fan delta deposited at the northern steep slope zone, with materials mainly from erosion of the Mesozoic volcanic rocks on Bonan Low uplifts. Weathering and dissolution of the volcanic material resulted in high contents of plagioclase and clay minerals of kaolinite and chlorite in the following diagenesis of Member 2. Further erosion of Bonan Low Uplift exposed larger areas of the Precambrian meta-granitoid, whose erosion offered large amounts of quartz and feldspar grains and low-clay-mineral content in the sandy conglomerates of Member 1.

5.4. Exploration Potential of the Kongdian Formation in Huanghekou Sag

The discussion above suggests the Kongdian Formation deposited in the steep slope zone of Bonan Low Uplift is characterized by internal subdivision with inhomogeneous mineral composition. Unlike Member 2 and Member 3 of the Kongdian Formation, drilled by borehole BZ28-A and BZ34-A in the east and Central Structural Belt, cores C1 and C2 represent Member 1 of the Kongdian Formation, with dominant quartz and low clay contents. The high contents of brittle minerals enhanced the development of the tectonic fissures, which significantly improve the permeability of the sandy conglomerate sequence under the background of intense fault activities. It is supposed that the sandy conglomerates of Member 1 of the Kongdian Formation in the steep slope zone of Huanghekou Sag might be of significant potential for oil and gas exploration.

6. Conclusions

In this study, we reported one U-Pb age of vein calcite and 235 detrital zircon ages of sandy conglomerate samples from borehole BZ26-A and conducted XRD analysis on samples from both BZ26-A and two well-understood boreholes of BZ28-A and BZ34-A. Based on the data and comparison, we draw two key conclusions:

(1) The absolute age of 47 ± 14.3 Ma of the vein calcite and the 103.5 ± 2.5 Ma of the youngest detrital zircon age suggest the thick sandy conglomerate sequence (~160 m in thickness) accumulated in the steep slope zone of Huanghekou Sag is supposed to be part of the Kongdian Formation, similar to those developed in the BZ19-6 Structure.

(2) Lithostratigraphy of the Kongdian Formation in Huanghekou Sag could be subdivided into at least three lithostratigraphic members with Member 3 consisting of red to brown sediment, deposited in a hot and dry climate, while Member 2 and Member 1 were deposited in fan delta facies, with major sediments from Mesozoic volcanic rocks and Precambrian meta-granitoid exposed on Bonan Low Uplift, respectively. Member 1 shows significant potential for further exploration.

Supplementary Materials: The following supporting information can be downloaded at: https://www.mdpi.com/article/10.3390/en15113880/s1, Table S1: Dataset of calcite dating; Table S2: Dataset of detrital zircon dating and Table S3: Dataset of XRD analysis.

Author Contributions: Data curation, L.H.; Investigation, W.W. and L.H.; Supervision, X.Y. and H.Z.; Writing—original draft, W.W.; Writing—review and editing, X.Y. and H.Z. All authors have read and agreed to the published version of the manuscript.

Funding: This study was funded by the National Natural Science Foundation of China (No. 42002131).

Institutional Review Board Statement: Not applicable.

Informed Consent Statement: Not applicable.

Data Availability Statement: The data are available in the Supplementary Materials.

Acknowledgments: This study was jointly supported by the National Natural Science Foundation of China (No. 42002131), the Fundamental Research Funds for the Central Universities; China University of Geosciences (Wuhan), (CUG2106328). We thank Zhixin Zhao with the Radiogenic Isotope Facility, the University of Queensland, for assistance with sample processing and in situ U-Pb analysis of calcite and data reduction. This manuscript benefited from detailed reviews by the reviewers.

Conflicts of Interest: The authors declare no conflict of interest.

References

1. Shi, H.S.; Wang, Q.B.; Wang, J.; Liu, X.J.; Feng, C.; Hao, Y.W.; Pang, W.J. Discovery and exploration significance of large condensate gas fields in BZ19-6 structure in deep Bozhong sag. *China Pet. Explor.* **2019**, *24*, 36–45, (In Chinese with English abstract).
2. Du, X.F.; Wang, Q.M.; Zhao, M.; Liu, X.J. Sedimentary characteristics and mechanism analysis of a large-scale fan delta system in the Paleocene Kongdian Formation, Southwestern Bohai Sea, China. *Interpretation* **2020**, *8*, SF81–SF94. [CrossRef]
3. Li, C.; Zhang, J.; Song, M.; Shi, C.; Shi, N.; Liu, T. Fine division and correlation of glutenite sedimentary periods based on sedimentary facies in version-A case study from the Paleogene strata of upper Es4 in the Yanjia Area, Dongying Depression. *Acta Geol. Sin.* **2011**, *85*, 1008–1018, (In Chinese with English abstract).
4. Cao, J.; Luo, J.; Ma, M.; Sheng, W.; Mao, Q.; Yang, T.; Di, W.; Song, K. Influence mechanism of micro-heterogeneity on conglomerate reservoir densification: A case study of Upper Permian Wutonggou Formation in DN8 Area of Dongdaohaizi Sag, Junggar Basin. *Earth Sci.* **2021**, *46*, 3435–3452.
5. Coogan, L.A.; Parrish, R.R.; Roberts, N.M.W. Early hydrothermal carbon uptake by the upper oceanic crust: Insight from in situ U-Pb dating. *Geology* **2016**, *44*, 147–150. [CrossRef]
6. Hansman, R.J.; Albert, R.; Gerdes, A.; Ring, U. Absolute ages of multiple generations of brittle structures by U-Pb dating of calcite. *Geology* **2018**, *46*, 207–210. [CrossRef]
7. Su, A.; Chen, H.H.; Feng, Y.X.; Zhao, J.X.; Nguyen, A.D.; Wang, Z.C.; Long, X.P. Dating and characterizing primary gas accumulation in Precambrian dolomite reservoirs, Central Sichuan Basin, China: Insights from pyrobitumen Re-Os and dolomite U-Pb geochronology. *Precambrian Res.* **2020**, *350*, 105897. [CrossRef]
8. Fedo, C.M.; Sircombe, K.; Rainbird, R. Detrital zircon analysis of the sedimentary record. *Rev. Mineral. Geochem.* **2003**, *53*, 277–303. [CrossRef]
9. Anderson, T. Detrital zircons as tracers of sedimentary provenance: Limiting conditions from statistics and numerical simulation. *Chem. Geol.* **2005**, *216*, 249–270. [CrossRef]
10. Gehrels, G. Detrital zircon U-Pb geochronology: Current methods and new opportunities. In *Tectonics of Sedimentary Basins: Recent Advances*; Busby, C., Azor, A., Eds.; Blackwell Publishing: Chichester, UK, 2012; pp. 45–62.
11. Qi, J.; Yang, Q. Cenozoic structural deformation and dynamic processes of the Bohai Bay basin province, China. *Mar. Pet. Geol.* **2010**, *27*, 757–771. [CrossRef]
12. Feng, Y.; Jiang, S.; Hu, S.; Li, S.; Lin, C.; Xie, X. Sequence stratigraphy and importance of syndepositional structural slope-break for architecture of Paleogene syn-rift lacustrine strata, Bohai Bay Basin, E. *China. Mar. Pet. Geol.* **2016**, *69*, 183–204. [CrossRef]
13. Qiu, N.; Zuo, Y.H.; Xu, W.; Li, W.Z.; Chang, J.; Zhu, C.Q. Meso-Cenozoic lithosphere thinning in the eastern North China Craton: Evidence from thermal history of the Bohai Bay Basin, North China. *J. Geol.* **2016**, *124*, 195–219. [CrossRef]
14. Li, J.P.; Zhou, X.H.; Liu, S.L.; Zhang, Y.H. On the distribution of the Kongdian Formation in the Bohai area with special reference to its bearings on oil exploration. *J. Stratigr.* **2010**, *34*, 89–96. (In Chinese with English abstract)
15. Yang, H.; Qian, G.; Xu, C.; Gao, Y.; Kang, R. Sandstone distribution and reservoir characteristics of Shahejie Formation in Huanghekou Sag, Bohai Bay Basin. *Earth Sci.* **2022**. Available online: https://kns.cnki.net/kcms/detail/42.1874.p.20220228.1831.006.html (accessed on 2 March 2022). (In Chinese with English abstract)
16. Li, H.; Wang, Q.B.; Pang, X.Q.; Dai, L.M.; Liu, X.J. Fracture generation and reservoir evaluation of tight glutenite reservoir: A case study of second member of Shahejie Formation in Huanghekou Depression. *Geol. Sci. Technol. Inf.* **2019**, *38*, 176–185. (In Chinese with English abstract)
17. Nafi, M.; Fei, Q.; Yang, X.H. Type of sandstone and source of carbonate cement in the Kongdian Formation (Upper Part), South Slope of the Dongying Depression, East China. *J. Appl. Sci.* **2004**, *4*, 235–241. [CrossRef]
18. Paton, C.; Hellstrom, J.; Paul, B.; Woodhead, J.; Hergt, J. Iolite: Freeware for the visualization and processing of mass spectrometric data. *J. Anal. Atmos. Spectrom.* **2011**, *26*, 2508–2518. [CrossRef]
19. Vermeesch, P. Isoplot R: A free and open toolbox for geochronology. *Geosci. Front.* **2018**, *9*, 1479–1493. [CrossRef]
20. Wiedenbeck, M.; Alle, P.; Corfu, F.; Griffin, W.L.; Meier, M.; Oberli, F.; Vonquadt, A.; Roddick, J.C.; Speigel, W. Three natural zircon standards for U-Th-Pb, Lu-Hf, trace element, and REE analyses. *Geostand. Geoanalytical Res.* **1995**, *19*, 1–23. [CrossRef]
21. Sláma, J.; Kosler, J.; Condon, D.J.; Crowley, J.L.; Gerdes, A.; Hanchar, J.M.; Horstwood, M.S.A.; Morris, G.A.; Nasdala, L.; Norberg, N.; et al. Plesovice zircon—A new natural reference material for U–Pb and Hf isotopic microanalysis. *Chem. Geol.* **2008**, *249*, 1–35. [CrossRef]
22. Vermeesch, P. On the visualisation of detrital age distributions. *Chem. Geol.* **2012**, *312–313*, 190–194. [CrossRef]
23. Zhai, M.G.; Santosh, M. The early Precambrian odyssey of the North China Craton: A synoptic overview. *Gondwana Res.* **2011**, *20*, 6–25. [CrossRef]
24. Zhao, G.; Cawood, P.A.; Li, S.; Wilde, S.A.; Sun, M.; Zhang, J.; He, Y.; Yin, C. Amalgamation of the North China Craton: Key issues and discussion. *Precambrian Res.* **2012**, *222*, 55–76. [CrossRef]
25. Zhai, Y.; Gao, S.; Zeng, Q.; Chu, S. Geochronology, geochemistry and Hf isotope of the Late Mesozoic granitoids from the Lushi polymetal mineralization area: Implication for the destruction of Southern North China Craton. *J. Earth Sci.* **2020**, *31*, 313–329. [CrossRef]
26. Meng, F.; Tian, Y.; Kerr, A.C.; Wang, Q.; Chen, Y.; Zhou, Y.; Yang, F. Neoarchean reworking of Mesoarchean and Paleoarchean crust (3.4 ~ 3.0 Ga) within the North China Craton: Constraints from zircon U-Pb geochronology and Lu-Hf isotopes from the basement of the Bohai Bay Basin. *Precambrian Res.* **2022**, *369*, 106497. [CrossRef]

27. Deng, J.; Su, S.; Niu, Y.; Liu, C.; Zhao, G.; Zhao, X.; Wu, Z. A possible model for the lithospheric thinning of North China Craton: Evidence from the Yanshanian (Jura-Cretaceous) magmatism and tectonism. *Lithos* **2007**, *96*, 22–35. [CrossRef]
28. Wang, Y.; Sun, L.; Zhou, L.; Xie, Y. Discussion on the relationship between the Yanshanian Movement and cratonic destruction in North China. *Sci. China Earth Sci.* **2018**, *61*, 499–514. [CrossRef]
29. Pang, X.J.; Wang, Q.B.; Xie, T.; Zhao, M.; Feng, C. Paleogene provenance and its control on high-quality reservoir in the northern margin of Huanghekou Sag. *Lithol. Reserv.* **2020**, *32*, 1–13. (In Chinese with English abstract)

Article

Modeling and Numerical Simulation of the Inlet Velocity on Oil–Water Two-Phase Vapor Separation Efficiency by the Hydrocyclone

Shuai Zhao [1,2], Jipeng Sun [2], Shuli Wang [2] and Zhihui Sun [1,*]

[1] School of Mines, China University of Mining and Technology, Xuzhou 221116, China; zhaoshuai6074@cumt.edu.cn
[2] Haiwang Hydrocyclone Co., Ltd., Postdoctoral Innovation Research Base, Weihai 264204, China; jipeng.sun@wh-hw.com (J.S.); shuli.wang@wh-hw.com (S.W.)
*   Correspondence: zhsun@cumt.edu.cn

Citation: Zhao, S.; Sun, J.; Wang, S.; Sun, Z. Modeling and Numerical Simulation of the Inlet Velocity on Oil–Water Two-Phase Vapor Separation Efficiency by the Hydrocyclone. *Energies* **2022**, *15*, 4900. https://doi.org/10.3390/en15134900

Academic Editors: Gan Feng, Qingxiang Meng, Gan Li and Fei Wu

Received: 30 May 2022
Accepted: 30 June 2022
Published: 4 July 2022

Publisher's Note: MDPI stays neutral with regard to jurisdictional claims in published maps and institutional affiliations.

Abstract: The density of tar vapor and water vapor produced by coal pyrolysis is different. Different centrifugal forces will be generated when they flow through the hydrocyclone. The water vapor and tar vapor are divided into inner and outer layers. According to this phenomenon, the moisture in the tar can be removed. In this paper, a Eulerian gas–liquid two-phase flow model is established by numerical simulation to study the effect of inlet velocity on the separation effect of a designed hydrocyclone (split ratio 0.2). The results show that the inlet velocity and moisture content have an influence on the volume distribution characteristics, tangential velocity, axial velocity, pressure drop distribution, and separation efficiency of tar vapor and water vapor in the hydrocyclone. When the inlet velocity increases from 2.0 to 12.0 m/s, the central swirl intensity increases, and the negative pressure sweep range at the overflow outlet increases. The axial velocity increased from 2.8 to 14.9 m/s, tar vapor content at the overflow outlet decreased from 74% to 37%, and at the underflow outlet increased from 89% to 92%. When the moisture content is lower than 10%, the hydrocyclone with the split ratio of 0.20 is no longer suitable for the separation of oil–water two-phase vapor. However, when the water content is higher than 20%, the purity of tar vapor at the underflow outlet can reach 92%, and the overflow outlet needs multistage separation to realize tar purification.

Keywords: pressure-controlled pyrolysis; oil–water separation; centrifugal force; two-phase vapor separation; cyclone separation

1. Introduction

Rich coal and less oil are the typical features of the Chinese energy structure. In recent years, the rising dependence on foreign crude oil has stimulated the progress of tar extraction from coal [1]. The pressure-controlled pyrolysis technology can extract tar from the abundant coal resources, and obtain important chemical raw materials, such as anthracene, naphthalene, benzene, and toluene. It is not only an important support in the field of the national energy and chemical industry [2,3], but also an important direction of coal classified utilization [4]. For a long time, the research on direct coal liquefaction has focused on the improvement of the catalytic hydrogenation process and equipment [5,6], catalyst selection and preparation [7,8], tar composition, and content optimization [9–11]. However, the water content of tar after coal pyrolysis reaches 10~30% and contains more sulfur ion and chloride ion salts. The presence of water leads to ion ionization and accelerates the corrosion of equipment.

When the temperature decreases, the viscosity of the tar will increase, and the tar will gradually become a non-Newtonian fluid, or even consolidate, which will bring difficulties to the oil–water separation process. The researchers tried to separate tar and water by different methods, such as sedimentation [12], centrifugation [13], heating [14], infiltration [15], and demulsification [16]. Gai et al. proposed the method of oil–water separation during coal pyrolysis by tar distillation decomposition. When the boiling point of the fraction is 120~140 °C and the volume ratio of the oil–water system is 1:4 at room temperature, the oil content of water can be reduced to less than 300 mg L^{-1} [17], but this method is greatly affected by temperature and is only suitable for less oil content. Li et al. established the kinetic model by using Smoluchowski aggregation theory and Stokes Einstein collision theory. Based on the heating and demulsification dehydration process, they predicted the effects of temperature, viscosity, time, and water content on tar dehydration efficiency [18]. Zelenskiy et al. proposed adding a demulsifier in the tar and water system to strengthen the process of gravity sedimentation, to separate the water in tar. The results show that the water content is reduced from 8~15% to 2~4% [19]. Although this process can remove most of the water from the tar, the demulsifier will dissolve in the tar and cause the secondary pollution.

To sum up, it is necessary to consider a high-speed and efficient separation device to realize the industrial process of oil–water separation. The cyclone separator is an efficient separation device based on medium density, which is widely used in the field of the petroleum and chemical industry. Wang et al., based on the Reynolds stress model (RSM) of the computational fluid dynamics (CFD) software, simulated the de-oiling effect of hydrocyclone with four different single-inlet structures. Within this research, the separation efficiency of the helical inlet hydrocyclone is the most ideal [20]. Sattar et al. thought that it took too long for electrical resistance tomography to generate fault maps, and proposed a new strategy based on the physics and simple logic behind the measurement to monitor the separation in the inline cyclone separator. The results show that the new algorithm is 104 times faster than the electrical resistance tomography in the case of inline eddy current separation [21]. Sattar et al. studied the influence of inlet conditions on the shape and size of the gas core in an inline swirl separator and proposed that the gas core diameter should be measured with the smallest possible invasion to predict the separation behavior and control the process in the best way [22]. The numerical modeling method can eliminate the influence of intrusion behavior on the performance of the hydrocyclone and obtain the strength and shape of the central swirl. Hanane et al. proposed a hybrid IBM-LES method and applied it to simulate laminar pipe flow for the first time. The results show that the method can control the fluid–solid interaction and converge to the exact solution with second-order accuracy [23]. Liu et al. designed a particle size reconstituted hydrocyclone and studied the effects of split ratio and oil content on the separation performance. The results show that the particle size reconstituted hydrocyclone can realize the separation of crude oil and water [24]. Jia et al. used the CFD–PBM numerical simulation method to study the separation mechanism of the hydrocyclone from different scales and showed that the oil–water separation effect was better when the split ratio was 0.2 [25].

After the low-rank coal was pyrolyzed by the pressure-controlled technology, the products flowed out of the high-temperature fluidized bed. At that time, the light fraction, water, and pyrolysis gas are all in the form of high-temperature vapor, with low viscosity and strong rheological characteristics. In this process, the light fraction and water are atomized at high temperature and have different densities. The cyclone separation process can be used to remove the water in the tar according to the difference of centrifugal force caused by the different density of light fractions and water vapor. In this paper, the effect of the hydrocyclone on separating oil–water two-phase vapor is studied by modeling and numerical simulation. According to the previous research results, the overflow split ratio of the hydrocyclone was selected as 0.2. The influence of inlet flow velocity on the stratification property of oil–water two-phase vapor flowing into the hydrocyclone under different water cut conditions is analyzed. The flow field distribution characteristics, phase distribution law, and pressure distribution characteristics of oil–water two-phase vapor in the hydrocyclone are explored, and the separation efficiency of the hydrocyclone with different inlet flow velocity and water content is revealed. This study aims to analyze the influence of process parameters on the separation efficiency of the finalized hydrocyclone, optimize the inlet flow corresponding to different water contents, and improve the separation efficiency of two-phase vapor, to optimize the structure of the hydrocyclone, promote the application field of the hydrocyclone, and provide guidance and help for the hydrocyclone separation technology of oil–water two-phase vapor.

2. Establish the Numerical Model

2.1. The Operation Process

The pressure-controlled pyrolysis process of low-rank coal was carried out with nitrogen as the protection and displacement medium, as shown in Figure 1. Nitrogen was input to the booster through the pressure relief valve. After the pressurization process is completed, it was stored in the air storage tank at the bottom of the booster. The pyrolysis process of low-rank coal was completed in a fluidized bed. High-pressure nitrogen was used to purge the fluidized bed, and the pyrolysis gas, tar vapor, and water vapor were purged into the cyclone separator to complete the oil–water separation process.

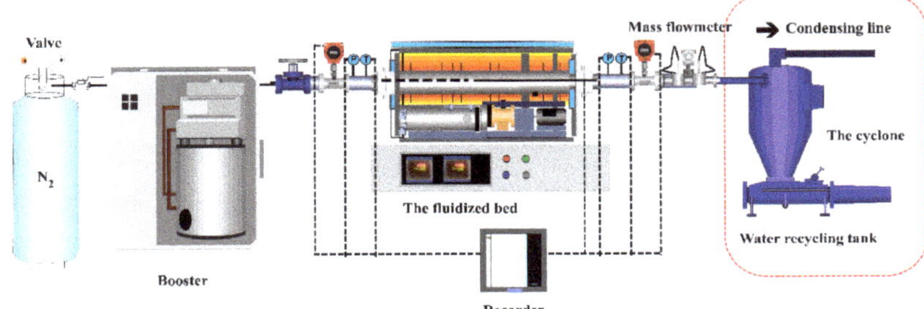

Figure 1. Low-rank coal pyrolysis and product separation system.

2.2. Geometric Model

The tangential inlet single-cone cyclone model was selected, and the structural parameters are shown in Table 1.

Table 1. The structural parameters of the cyclone.

Attribute	Symbol	Value	Unit
Inlet cross-sectional area	A_1	453.00	mm^2
Overflow pipe diameter	D_1	10.70	mm
Length of overflow pipe	L_1	25.00	mm
Depth of overflow pipe	H_1	12.50	mm
Spacing between overflow pipe and swirl chamber	L_2	6.25	mm
Swirl chamber diameter	D_2	60.00	mm
Swirl chamber height	H_2	75.00	mm
Cone length	L_3	276	mm
Cone angle	α	11.75	°
Underflow outlet diameter	D_3	12.5	mm

The fluent preprocessor was used to divide the hexahedral mesh, as shown in Figure 2. The grid cell size was 1.0 mm, and grid encryption was carried out at joints, corners, entrances, and exits [26]. The pressure loss along the way was selected for grid independence verification. The number of grids was 327,742, 237,682, and 124,578, respectively, and the pressure drop was 0.182, 0.176, and 0.137 kPa, respectively. In this paper, the model with 237,682 grids was selected to complete the simulation. The type of boundary layer was defined as no slip boundary, and the thickness was 1.0 mm.

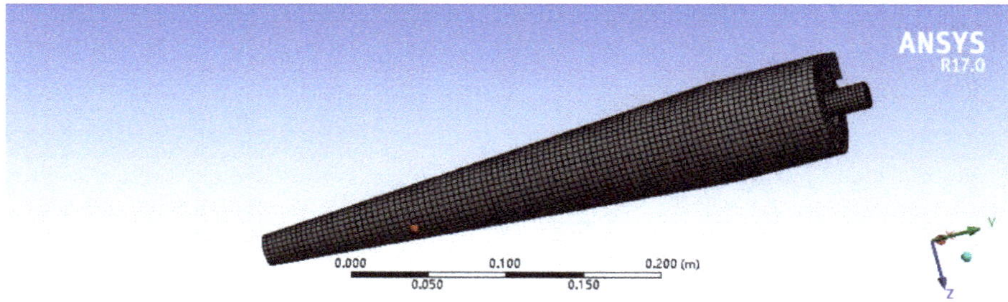

Figure 2. Grid division of the cyclone.

2.3. Governing Equations

The volume distribution, turbulence intensity, tangential velocity, and axial velocity of tar vapor and water vapor in the hydrocyclone follow the conservation law of mass and energy. The moisture content was between 10% and 30%, which meets the selection characteristics of a dispersed phase volume ratio exceeding 10%. Therefore, the Euler method was selected for fluid analysis of oil–water two-phase vapor. The parameter characteristics are shown in Table 2.

Table 2. The characteristic parameters of oil–water two-phase vapor.

Attribute	Symbol	Value	Unit
Tar vapor density	ρ_1	4.84	kg/m^3
Tar vapor heat transfer coefficient	k_1	0.0178	$W/m^2 \cdot K$
Tar vapor viscosity	μ_1	6.75×10^{-6}	$N \cdot s/m^2$
Water vapor density	ρ_2	0.5542	kg/m^3
Water vapor heat transfer coefficient	k_2	0.0261	$W/m^2 \cdot K$
Water vapor viscosity	μ_2	1.34×10^{-5}	$N \cdot s/m^2$
Convection term C_1	C_1	1.44	-
Convection term C_2	C_2	1.92	-
Tar vapor particle size	d	0.0004	m
Moisture content		10%; 20%; 30%; 40%	-
Inlet velocity	u	2; 4; 8; 12	m/s

The separation process of two-phase vapor in the cyclone presented an internal and external two-layer swirl; especially, the turbulence degree of the internal swirl was higher, and the anisotropy was more obvious [27,28]. The Reynolds stress model (RSM) considers the characteristics of rapid change of vortex, rotation, and tension in engineering turbulence calculation, and can solve the problem of anisotropy of the internal and external vortex [29].

Transport equation of each component in RSM:

$$\frac{\partial\left(\rho\overline{u_i'u_j'}\right)}{\partial t} + \frac{\partial\left(\rho u_k\overline{u_i'u_j'}\right)}{\partial x_k} = \frac{D\left(\rho\overline{u_i'u_j'}\right)}{Dt} = D_{i,j} + P_{i,j} + G_{i,j} + \Phi_{i,j} - \varepsilon_{i,j} + F_{i,j}$$

where $D_{i,j}$—the diffusion term of the turbulence model, $P_{i,j}$—the shear stress generation term, $G_{i,j}$—the buoyancy generation term, $\varphi_{i,j}$—pressure strain redistribution, $\varepsilon_{i,j}$—the viscous dissipation term, and $F_{i,j}$—the rotation generation term of two-phase system.

$$D_{i,j} = -\frac{\partial}{\partial t}\left[\rho\overline{u_i'u_j'u_k'} + \overline{P\left(\delta_{kj}u_i' + \delta_{ik}u_j'\right)} - \mu\frac{\partial}{\partial x_k}\left(\overline{u_i'u_j'}\right)\right]$$

$$D_{i,j} = -\frac{\partial}{\partial t}\left[\rho\overline{u_i'u_j'u_k'} + \overline{P\left(\delta_{kj}u_i' + \delta_{ik}u_j'\right)} - \mu\frac{\partial}{\partial x_k}\left(\overline{u_i'u_j'}\right)\right]$$

$$P_{i,j} = -\rho\left(\overline{u_i'u_k'}\frac{\partial \overline{u_j'}}{\partial x_k} + \overline{u_j'u_k'}\frac{\partial \overline{u_i'}}{\partial x_k}\right)$$

$$G_{i,j} = -\rho\beta\left(g_i\overline{u_j'\theta} + g_j\overline{u_i'\theta}\right)$$

$$\Phi_{i,j} = P\left(\overline{\frac{\partial u_j'}{\partial x_i} + \frac{\partial u_i'}{\partial x_j}}\right)$$

$$\varepsilon_{i,j} = 2\mu\frac{\overline{\partial u_i'\partial u_j'}}{\partial x_k\partial x_k}$$

$$F_{i,j} = -2\rho\Omega_x\left(\overline{u_j'u_m'}\varepsilon_{ikm} + \overline{u_i'u_m'}\varepsilon_{jkm}\right)$$

where β—coefficient of thermal expansion, m/K, Ω—vortex vector, θ—cone angle of cyclone, °, and g—gravitational acceleration, -9.81 m/s^2.

After calculation and simplification, the final RSM transport equation is as follows:

$$\frac{\partial\left(\rho\overline{u_i'u_j'}\right)}{\partial t} + \frac{\partial\left(\rho u_k\overline{u_i'u_j'}\right)}{\partial x_k} = \frac{\partial}{\partial x_k}\left(\frac{\mu_i}{\delta_{ik}}\frac{\overline{\partial u_i'u_j'}}{\partial x_k} + \mu\frac{\overline{\partial u_i'u_j'}}{\partial x_k}\right) - \rho\left(\overline{u_i'u_k'}\frac{\partial u_j}{\partial x_k} + \overline{u_j'u_k'}\frac{\partial u_i}{\partial x_k}\right) - C_1\rho\frac{\varepsilon}{k}\left(\overline{u_iu_j} - \frac{2}{3}k\delta_{i,j}\right) - C_2\left(P_{i,j} - \frac{1}{3}P_{kk}\delta_{i,j}\right) - \frac{2}{3}\rho\varepsilon\delta_{i,j}$$

where ρ—density, kg/m^3, P—pressure, Pa, K—thermal conductivity, W/(m^2·K), μ—Hydro dynamic viscosity, (N·s)/m^2, u_i—velocity vector, and C_1,C_2—convection coefficients of the k–ε equation.

3. Results and Discussion

3.1. Velocity Distribution

Velocity is an important factor affecting the separation efficiency of the cyclone, especially the tangential velocity, which determines the centrifugal force [30,31]. As shown in Figure 3, it is the velocity nephogram of tar vapor in the cyclone under different inlet velocities. The flow velocity was large at the overflow outlet, swirl chamber, and cone wall of the cyclone. That is because in the hydrocyclone, two layers of internal and external swirls with opposite directions were formed. The internal swirls are also called the forced vortex zone. The external swirling flow is also called the semi-free vortex zone. When the inlet velocity was 2.0 m/s, the hydrocyclone could not form the swirl or the forced vortex zone. When the inlet velocity was greater than 4.0 m/s, the vortex intensity in the forced vortex zone gradually increased, and the higher the inlet velocity, the smaller the radius of the forced vortex zone and the stronger the suction capacity. The main component of the forced vortex zone is water vapor. The density of water vapor was smaller than that of tar vapor, and the centrifugal force was lower. However, due to the suction of high-speed fluid, the turbulence degree of water vapor was high, so the axial velocity was high. The tangential velocity in the center of the forced vortex zone was almost 0. The main component is tar vapor in the semi-free vortex zone. Due to the high density, the tar vapor also bears high centrifugal force, which leads to the small axial velocity of tar vapor and the large tangential velocity and has the characteristics of wall attachment.

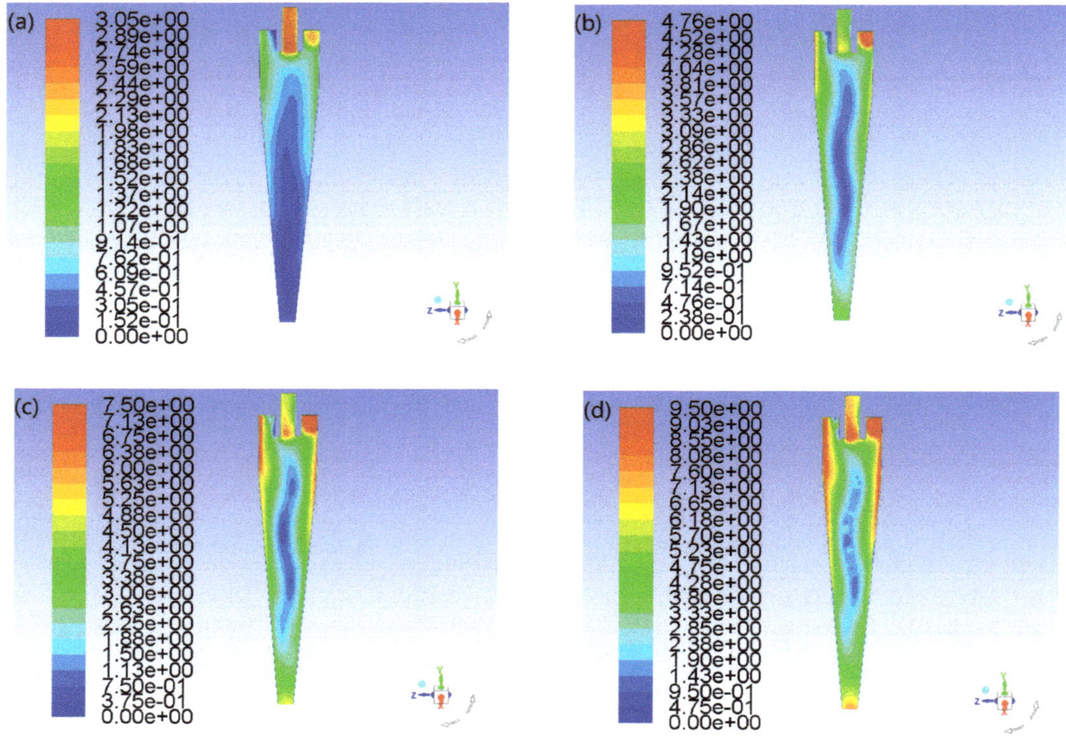

Figure 3. The velocity nephogram of tar vapor under different inlet velocities: (**a**) 2.0 m/s, (**b**) 4.0 m/s, (**c**) 8.0 m/s, and (**d**) 12.0 m/s.

The monitoring curve was set at the conical section and central axis of the cyclone, and the tangential velocity of the cyclone was asymmetrically distributed along the radial direction, as shown in Figure 4a. The flow velocity at the cone section wall was higher and increased with the increase of the inlet flow velocity. This is because when the moisture content is the same, the centrifugal force will increase with the increase of the inlet velocity, and then the tangential velocity will also increase [32]. The influence of the inlet velocity on the axial velocity is mainly reflected in the overflow outlet, as shown in Figure 4b. With the increase of the inlet velocity, the velocity at the underflow outlet and cone section showed little change, while the velocity at the overflow outlet increased a lot. This is because the outflow phase at the underflow outlet was mainly tar vapor with high density, and the flow velocity of tar vapor along the cone wall at the edge of the swirl was low. Even if the inlet velocity is increased, the rate of tar vapor wall-attached flow will not greatly increase.

At the same inlet velocity, the moisture content of two−phase vapor had a weak effect on the distribution of tangential and axial velocity. With the increase of moisture content, the tangential velocity decreased a little, while the axial velocity increased a little, as shown in Figure 4c,d. This is because the density of water vapor is lower than that of tar vapor. When the inlet velocity is constant, the increase of light phase content means that the content of tar vapor is relatively reduced. This led to the decrease of the average density of the two-phase system, so the centrifugal force decreased accordingly, and the tangential velocity also decreased. On the other hand, the negative pressure formed by axial swirl had a suction effect on the central fluid. With the increase of water vapor content, the central fluid density decreased relatively, the effect of suction was amplified, and the velocity of the axial overflow outlet increased. However, the flow velocity at the underflow outlet is mainly the wall-attached flow of tar vapor, so the moisture content had little effect on it.

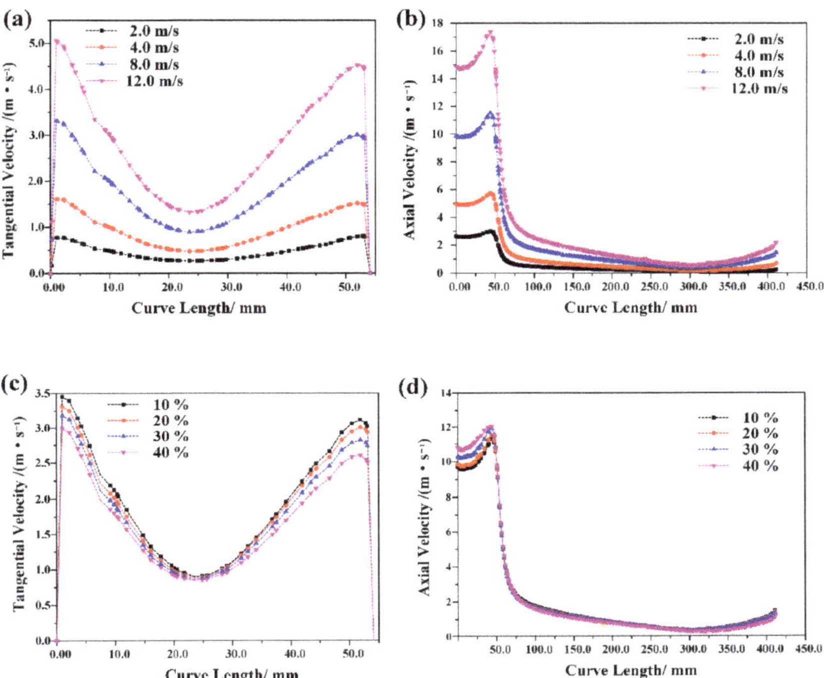

Figure 4. The velocity distribution of tar vapor. (**a**) Tangential velocity with water content of 20%, (**b**) axial velocity with water content of 20%, (**c**) tangential velocity when inlet velocity is 8.0 m/s, and (**d**) axial velocity when inlet velocity is 8.0 m/s.

3.2. The Partial Pressure of Tar Vapor Phase

As shown in Figure 5, when the water vapor content of the two-phase system was 20%, the tar stream and water vapor distribution occurred in the section of the cyclone. When the inlet velocity was 2.0 m/s, the intensity of the axial swirl in the center of the cyclone was low, and the suction effect was not obvious. The water vapor with lower density accumulated in the swirl chamber and flowed out from the overflow outlet, and only a small amount of tar vapor flowed on the wall. When the inlet velocity was higher than 4.0 m/s, the intensity of the central swirl increased and the suction effect on water vapor phase was more obvious. The aggregation degree of tar vapor in the outer swirl increased with the increase of the inlet velocity.

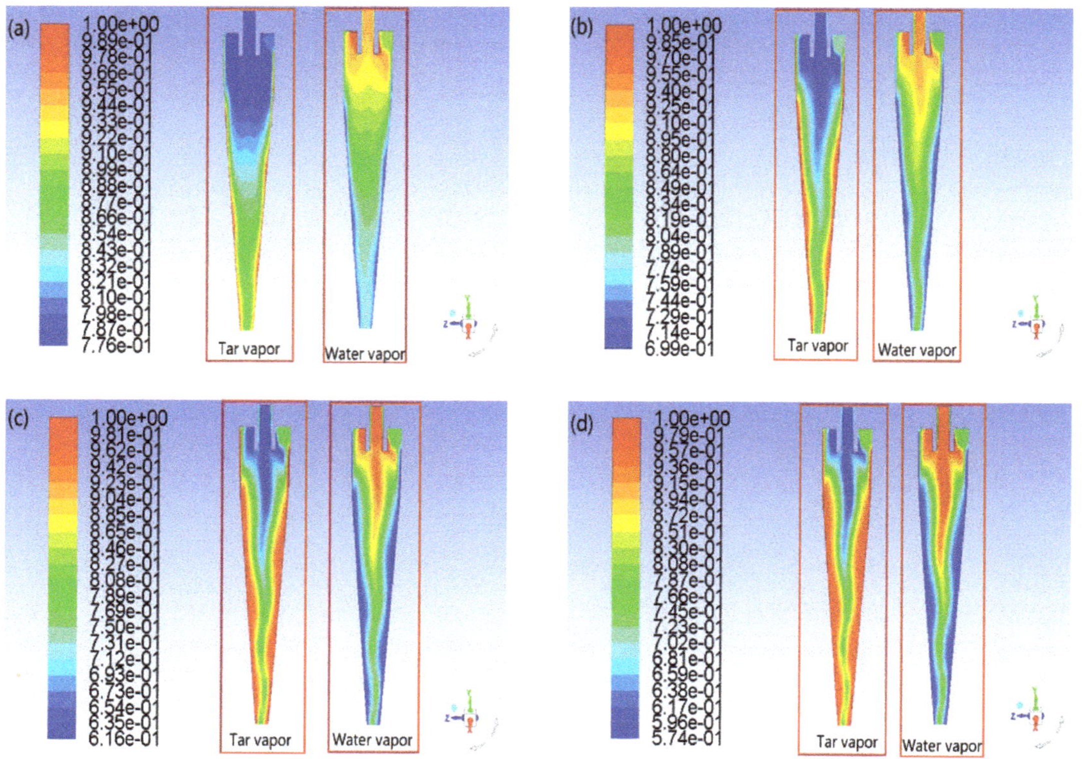

Figure 5. The phase distribution of tar vapor and water vapor in cyclone section with water content of 20%. (**a**) Inlet velocity 2.0 m/s, (**b**) inlet velocity 4.0 m/s, (**c**) inlet velocity 8.0 m/s, and (**d**) inlet velocity 12.0 m/s.

The partial pressure of tar vapor in the radial and axial direction of the cyclone also confirmed the above view. As shown in Figure 6a, when the moisture content of the two-phase system was 20% and the inlet velocity was 2.0 m/s, there was no obvious difference in the radial partial pressure of tar vapor. When the inlet velocity was greater than 4.0 m/s, the partial pressure of tar vapor increased in the radial direction, especially within 10 mm of the cone wall. The partial pressure of tar vapor in the center of the cyclone was low, and this trend was more obvious with the increase of the inlet velocity. When the inlet velocity was 8.0 m/s, as shown in Figure 6b, the content of water vapor also had a weak influence on the partial pressure of tar vapor. With the increase of the water vapor content, the radial partial pressure of tar vapor decreased, especially within 10 mm of the cone wall. This is mainly because the centrifugal force of tar vapor decreased when the inlet velocity was constant. The increased water vapor content led the average density to decrease in the two-phase system. The axial partial pressure of tar vapor in the cyclone also has similar characteristics. As shown in Figure 6c, when the moisture content of the two-phase system was 20% and the inlet velocity was 2.0 m/s, the axial partial pressure of tar vapor was only slightly different at the overflow outlet, swirl chamber, and cone section. With the increase of the inlet velocity, the partial pressure of tar vapor at the swirl chamber and cone section increased, and the partial pressure of tar vapor at the overflow outlet was almost without any change. The contact position between the overflow pipe and the swirl chamber presented negative pressure, which was caused by the suction effect of the axial swirl. This is beneficial to increase the axial velocity, reduce the tangential velocity in the center of the cyclone, and improve the separation efficiency. In addition, with the increase of the water vapor content, as shown in Figure 6d, there was no significant difference of pressure in the overflow outlet when the inlet velocity was 8.0 m/s. The pressure of the swirl chamber and the cone section showed a decreasing trend. Combined with the nephogram of tar vapor distribution, the increase of water vapor content led to the increase of the central cyclone degree, so the centrifugal force of tar vapor increased, and the content in the center of the cyclone decreased.

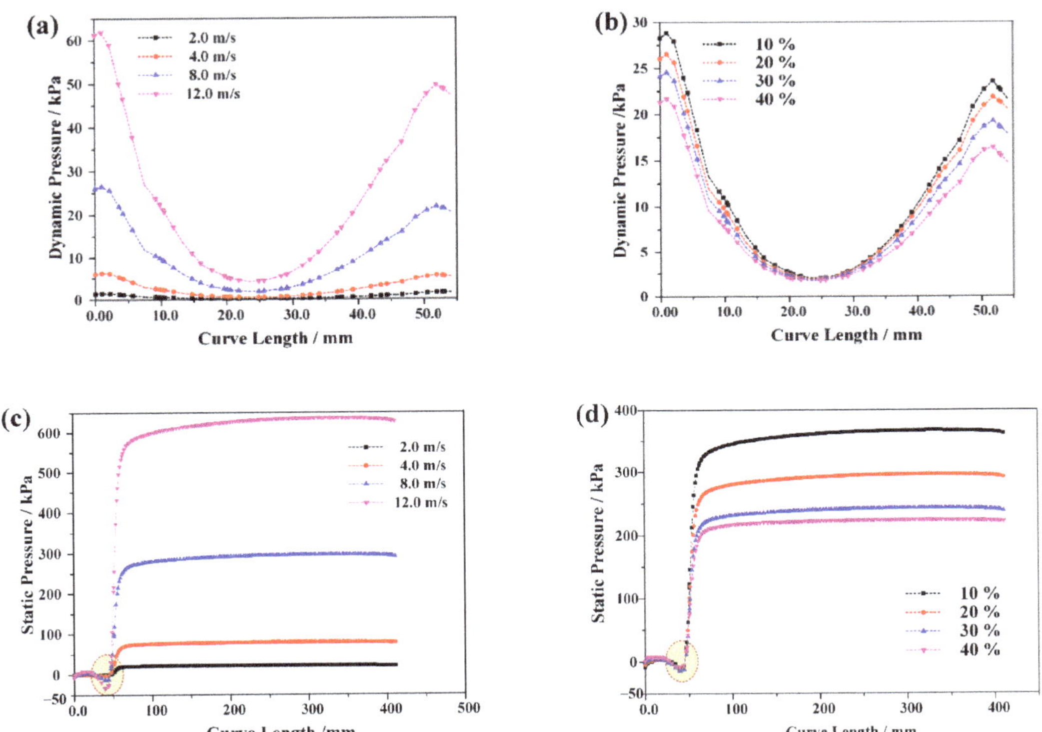

Figure 6. The partial pressure of tar vapor in radial and axial direction of cyclone. (**a**) Radial pressure with water content of 20%, (**b**) radial pressure when inlet velocity is 8.0 m/s, (**c**) axial pressure with water content of 20%, and (**d**) axial pressure when inlet velocity is 8.0 m/s.

3.3. Volume Distribution Characteristics of Tar Vapor

The streamline distribution characteristics of water vapor flowing in the cyclone are shown in Figure 7a, b. The density of water vapor was smaller than that of tar vapor, which was mainly distributed in the upper layer of the cyclone and flowed out through the overflow outlet. When the inlet velocity was within the range of 2.0~12.0 m/s, the flow area of water vapor was greatly affected by the inlet velocity. When the inlet velocity was 4.0 m/s, the water vapor flow area was in the upper part of the swirl chamber and the cone section. With the increase of inlet velocity, the entrainment of the internal swirl formed in the center of the cyclone was strengthened, and the occupied volume of water vapor in the swirl chamber was gradually reduced.

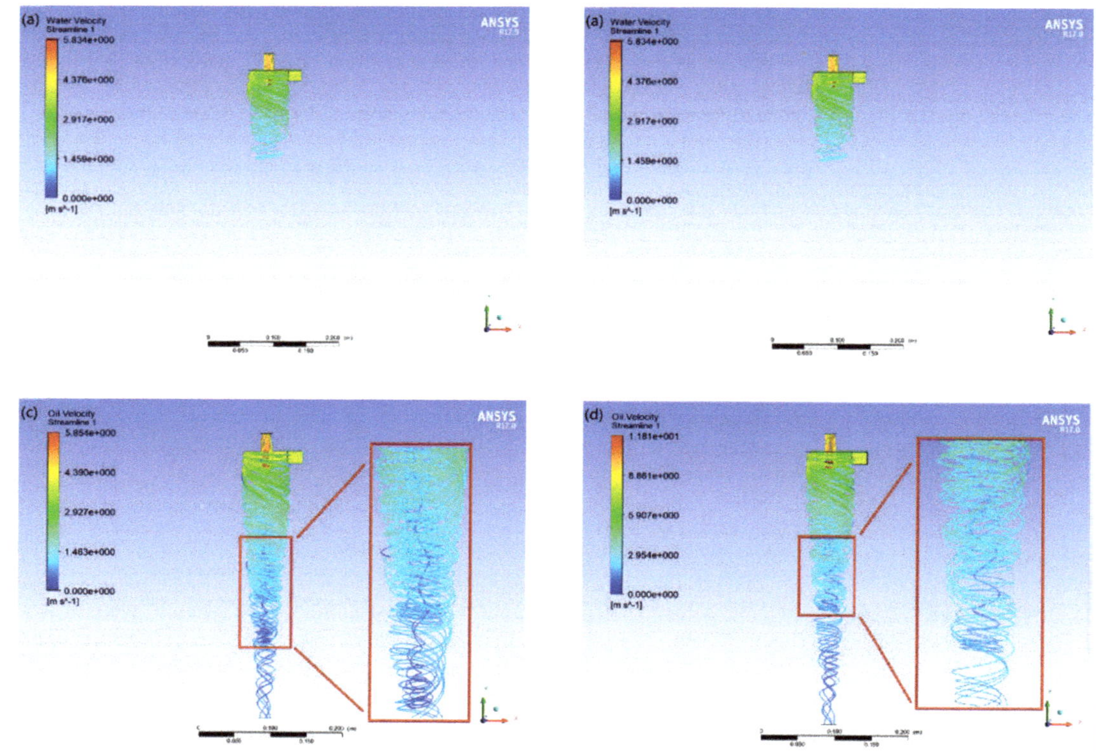

Figure 7. The streamline distribution of water vapor and tar vapor in cyclone. (**a**) Water vapor streamline when inlet velocity is 4.0 m/s, (**b**) water vapor streamline when inlet velocity is 8.0 m/s, (**c**) tar vapor streamline when inlet velocity is 4.0 m/s, and (**d**) tar vapor streamline when inlet velocity is 8.0 m/s.

The tar vapor with high density mainly flows out from the underflow outlet of the cyclone, but some tar vapor will still flow out from the overflow outlet due to the entrainment of the internal cyclone. With the increase of inlet velocity, the centrifugal force on tar vapor increased, so the entrainment effect of the central swirl on tar vapor gradually decreased, as shown in Figure 7c,d. The distribution of the tar vapor flow line in the internal swirl gradually decreased with the increase of the inlet velocity. This is mainly because the increase of the inlet velocity led to the increase of tangential velocity of the two-phase vapor in the swirl chamber. The centrifugal force of tar vapor was further increased, and more tar vapor was thrown to the outside of the swirl, and then flowed, attached to the wall [32].

The volume distribution of tar vapor in the radial direction of the cyclone is shown in Figure 8. The inlet velocity and moisture content had significant effects on the distribution of tar vapor. The tar vapor with a content greater than 85% was mainly distributed within 10 mm of the outer swirl. When the moisture content was 20%, as shown in Figure 8a, the content of tar vapor in the central swirl was 67~83%. With the increase of inlet velocity, the tar vapor content of the central swirl decreased, and the tar vapor content gradually increased within 10 mm to the wall. When the inlet velocity was 8.0 m/s, as shown in Figure 8b, with the increase of moisture content, the volume fraction difference of the tar vapor between the outside and the center of the swirl increased, which is more conducive to the separation of the two-phase vapor.

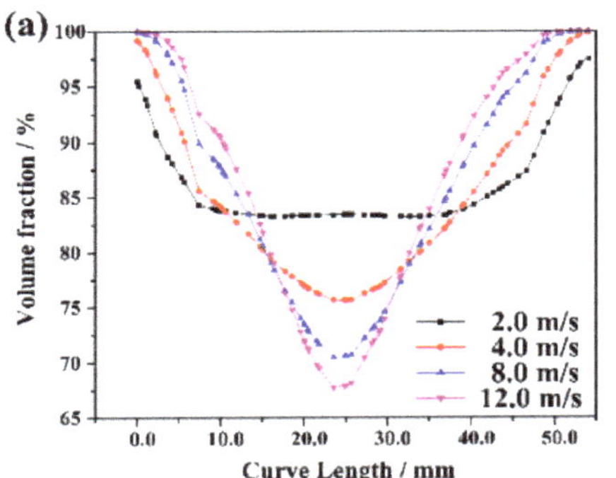
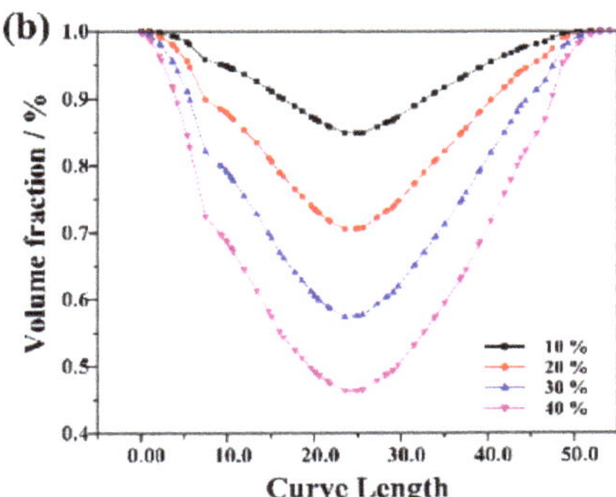

Figure 8. Volume distribution characteristics of tar vapor in the radial direction of the cyclone. (**a**) The water content is 20% and (**b**) the inlet velocity is 8.0 m/s.

The axial distribution difference of tar vapor is mainly reflected in the volume fraction flowing out from overflow and underflow outlet. As shown in Figure 9a, when the moisture content was 20%, the volume fraction of tar vapor at the overflow outlet was about 70~86.43%, and the volume fraction at the underflow outlet was 92.07%. The increase of inlet velocity will reduce the volume fraction of tar vapor at the overflow outlet, but it had little effect on the volume fraction of tar vapor at the underflow outlet. When the inlet velocity was 8.0 m/s and the moisture content was 10%, as shown in Figure 9b, the volume fraction of tar vapor at the overflow outlet was 74.69%. Under the condition of constant inlet velocity, with the increase of moisture content, the volume fraction of tar vapor at the overflow outlet gradually decreased to 37.54%; however, at the underflow outlet, it was almost unaffected, about 90%.

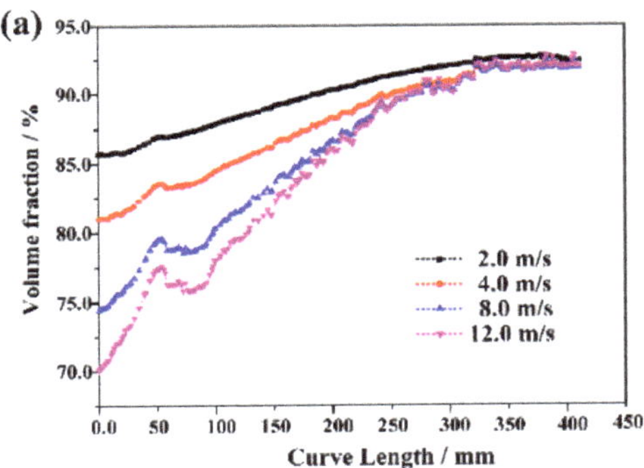
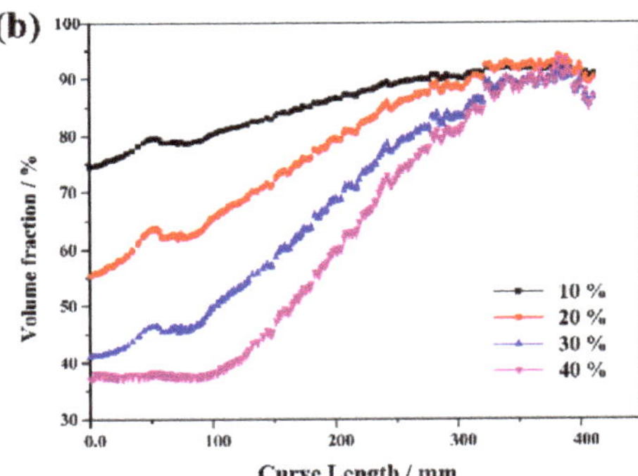

Figure 9. Volume distribution characteristics of tar vapor in the axial direction of the cyclone. (**a**) The water content is 20% and (**b**) the inlet velocity is 8.0 m/s.

The separation efficiency of the two-phase vapor is an important parameter to measure the working performance of the cyclone. As shown in Figure 10a, when the moisture content was 20%, the influence of inlet velocity on the separation efficiency of the cyclone was mainly reflected in the volume fraction of tar vapor at the overflow outlet. When the inlet velocity increased from 2.0 to 12.0 m/s, the volume fraction of tar vapor at the overflow outlet decreased from 74.27% to 47.76%, while the tar vapor volume fraction at the underflow outlet changed slightly, from 89.02% to 94.04%. In addition, the moisture content of the two-phase system also affects the separation efficiency of the cyclone, as shown in Figure 10b. When the inlet velocity was 8.0 m/s and the moisture content increased from 10% to 40%, the content of tar vapor at the overflow outlet decreased from 75.04% to 37.56%, and at the underflow outlet it decreased from 97.41% to 92.34%. The cyclone is not suitable for oil–water separation when the moisture content of the two-phase system is less than 10%. When the moisture content is higher than 20%, the two-phase system flowing out from the overflow outlet shall adopt the form of multi-stage cyclone combination to separate the oil–water of the two-phase vapor again.

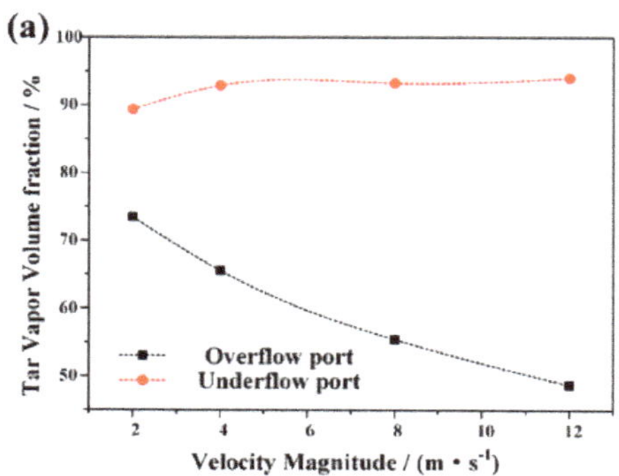
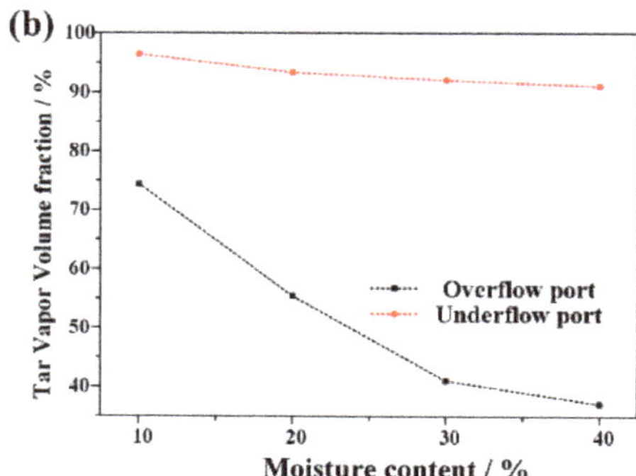

Figure 10. The separation efficiency of the cyclone. (**a**) At the overflow outlet and (**b**) at the under-flow outlet.

4. Conclusions

In this paper, the effect of inlet velocity on the separation efficiency of the designed hydrocyclone was studied by numerical simulation. A Eulerian gas–liquid two-phase flow model was established to investigate the variation trend of inlet velocity and moisture content on the volume distribution characteristics, radial pressure drop, tangential velocity, and axial velocity of tar vapor and water vapor in the designed hydrocyclone, with a split ratio of 0.2. When the water content was 20%, the inlet velocity increased from 2.0 to 12.0 m/s, and the separation efficiency of the cyclone also increased from 89.02% to 94.04%. However, the increase of water content will reduce the radial pressure gradient of the hydrocyclone, weaken the turbulence degree in the forced vortex area of the hydrocyclone, and then reduce the stratification effect. On the other hand, when the moisture content of the two-phase system was less than 10%, the cyclone was not suitable for oil–water separation. When the moisture content was higher than 20%, the two-phase system flowing out from the overflow outlet shall adopt the form of multistage cyclone combination to separate the oil–water of the two-phase vapor again.

Author Contributions: Formal analysis, writing—original draft preparation, S.Z.; writing—review and editing, Z.S., methodology, project administration, J.S. and S.W. All authors have read and agreed to the published version of the manuscript.

Funding: This research was funded by "Jiangsu Natural Science Youth Fund grant number BK20221133", "The Fundamental Research Funds for the Central Universities, grant number 2021QN1002", "China Postdoctoral Science Foundation, grant number 2021M693421", and "The 2021 Jiangsu Shuang chuang (Mass Innovation and Entrepreneurship) Talent Program, grant number JSSCBS20211238".

Acknowledgments: This study was supported by "Jiangsu Natural Science Youth Fund, grant number BK20221133", "The Fundamental Research Funds for the Central Universities, grant number 2021QN1002", "China Postdoctoral Science Foundation, grant number 2021M693421", and "The 2021 Jiangsu Shuangchuang (Mass Innovation and Entrepreneurship) Talent Program, grant number JSSCBS20211238". The authors also express their appreciation to the technical reviewers for their constructive comments.

Conflicts of Interest: The authors declare no conflict of interest.

References

1. Wu, Q.; Tu, K.; Zeng, Y.F.; Liu, S.Q. Discussion on the main problems and countermeasures for building an upgrade version of main energy (coal) industry in China. *J. China Coal Soc.* **2019**, *44*, 1625–1636.
2. Zou, C.; Chen, Y.; Kong, L.; Sun, F.; Shanshan, C.; Zhen, D. Underground coal gasification and its strategic significance to the development of natural gas industry in China. *Pet. Explor. Dev.* **2019**, *46*, 195–204. [CrossRef]
3. Yang, J.X.; Wei, W.J.; Qi, Y. Research progress on hyper-coal for efficient utilization. *J. China Coal Soc.* **2020**, *9*, 3301–3313.
4. Li, J.; Mao, X.; Zhong, J.; Liu, M. Carbon Nanotubes Catalyst Modified via NiMoP and Its Performance in the Hydrogenation of Direct Coal Liquefaction Oil. *Acta Pet. Sin.* **2019**, *35*, 1167–1174.
5. Liu, M.; Chen, G.F.; Wang, Y.G.; Zhao, P.; Qu, S.J. Conversion of sulphur and nitrogen compounds in hydrofining process of Baishihu coal liquefaction oil. *J. Fuel Chem. Technol.* **2019**, *47*, 870–875.
6. Liu, L.; Zhao, L.; Sun, Y.; Gao, S.; Jiang, M.; Jiang, M.; Rosso, D. Separation performance of hydrocyclones with medium rearrangement internals. *J. Environ. Chem. Eng.* **2021**, *9*, 105642. [CrossRef]
7. Shuai, Z.; Xiaoshu, L.; Qiang, L.; Youhong, S. Thermal-fluid coupling analysis of oil shale pyrolysis and displacement by heat-carrying supercritical carbon dioxide. *Chem. Eng. J.* **2020**, *394*, 125037. [CrossRef]
8. Tong, D.; Zhang, Q.; Liu, F.; Geng, G.; Zheng, Y.; Xue, T.; Hong, C.; Wu, R.; Qin, Y.; He, K. Current Emissions and Future Mitigation Pathways of Coal-Fired Power Plants in China from 2010 to 2030. *Environ. Sci. Technol.* **2018**, *52*, 12905–12914. [CrossRef]
9. Apicella, B.; Russo, C.; Cerciello, F.; Stanzione, F.; Ciajolo, A.; Scherer, V.; Senneca, O. Insights on the role of primary and secondary tar reactions in soot inception during fast pyrolysis of coal. *Fuel* **2020**, *275*, 11795. [CrossRef]
10. Huang, X.; Cheng, D.G.; Chen, F.; Zhan, X. The decomposition of aromatic hydrocarbons during coal pyrolysis in hydrogen plasma: A density functional theory study. *Int. J. Hydrog. Energy* **2012**, *37*, 18040–18049. [CrossRef]
11. Feng, J.; Xue, X.; Li, X.; Li, W.; Guo, X.; Liu, K. Products analysis of Shendong long-flame coal hydropyrolysis with iron-based catalysts. *Fuel Process. Technol.* **2015**, *130*, 96–100. [CrossRef]
12. Tian, Y.; Feng, J.; Cai, Z.; Chao, J.; Zhang, D.; Cui, Y.; Chen, F. Dodecyl Mercaptan Functionalized Copper Mesh for Water Repellence and Oil-water Separation. *J. Bionic Eng.* **2021**, *18*, 887–899. [CrossRef]
13. Zhang, X.; Liu, C.; Yang, J.; Huang, X.J.; Xu, Z.K. Wettability Switchable Membranes for Separating Both Oil-in-water and water-in-oil emulsions. *J. Membr. Sci.* **2021**, *624*, 118976. [CrossRef]
14. Hafsi, Z.; Elaoud, S.; Mishra, M.; Wada, I. Numerical Study of Droplets Coalescence in an Oil-Water Separator. In Proceedings of the International Conference on Advances in Mechanical Engineering and Mechanics; Springer: Cham, Switzerland, 2021; pp. 449–454.
15. Kovaleva, L.; Zinnatullin, R.; Musin, A.; Gabdrafikov, A.; Sultanguzhin, R.; Kireev, V. Influence of radio-frequency and microwave electromagnetic treatment on water-in-oil emulsion separation. *Colloids Surf. A Physicochem. Eng. Asp.* **2021**, *614*, 126081. [CrossRef]
16. Feng, Y.; Yu, B.; Zhang, X.; Gong, H.; Liu, Y. Numerical simulation on the effect of combining centrifugation, electric field and temperature on two-phase separation. *Chem. Eng. Process.* **2020**, *148*, 107803. [CrossRef]
17. Gai, H.; Zhang, X.; Chen, S.; Wang, C.; Xiao, M.; Huang, T.; Wang, J.; Song, H. An improved tar–water separation process of low–rank coal conversion wastewater for increasing the tar yield and reducing the oil content in wastewater. *Chem. Eng. J.* **2020**, *333*, 123229. [CrossRef]
18. Li, H.; Li, D.; Yang, X.; Wang, J.; Li, B.; Li, W.; Zhao, L. Kinetics for dehydration of coal tar. *Chem. Eng. J.* **2011**, *9*, 57–60.
19. Zelenskiy, V.V.; Fillipov, V.L.; Bannikov, L.P.; Karchakova, V.V. Intensifying the separation of water-tar emulsions. *Coke Chem.* **2014**, *57*, 163–166. [CrossRef]
20. Wang, Y.; Zhao, L.; Wang, Y.; Baorui, X.; Minghu, J. Contrastive Analysis on Numerical Simulation of Oil-water Hydrocyclone with Various Inlet Forms. *Fluid Mach.* **2017**, *45*, 22–27.
21. Sattar, M.A.; Garcia, M.M.; Portela, L.M.; Babout, L. A Fast Electrical Resistivity-Based Algorithm to Measure and Visualize Two-Phase Swirling Flows. *Sensors* **2022**, *22*, 1834. [CrossRef]
22. Sattar, M.A.; Garcia, M.M.; Banasiak, R.; Portela, L.M.; Babout, L. Electrical Resistance Tomography for Control Applications: Quantitative Study of the Gas-Liquid Distribution inside A Cyclone. *Sensors* **2020**, *20*, 6069. [CrossRef] [PubMed]
23. Amani, H.; Zamansky, R.; Climent, E.; Legendre, D. Stochastic wall model for turbulent pipe flow using Immersed Boundary Method and Large Eddy Simulation. *Comput. Fluids* **2022**, *239*, 105419. [CrossRef]
24. Lu, Y.; Ma, J.; He, Y.; Zhao, Y.; Sun, C.; Yin, P.; Zhang, S. Study on the Oil-Water Separation Performance of the Droplet Size Reconstruction Hydrocyclone. *China Pet. Mach.* **2020**, *48*, 90–97.
25. Jia, P.; Chen, J.; Cai, X.; Kong, L.; Wang, C.; Shang, C.; Zhang, M.; Shi, Y. Study on Oil-Water Separation Characteristics of Hydrocyclone Based on CFD-PBM Numerical Simulation. *J. Petrochem. Univ.* **2021**, *34*, 58–65.
26. Nunes, S.A.; Magalhães, H.L.; Gomez, R.S.; Vilela, A.F.; Figueiredo, M.J.; Santos, R.S.; Rolim, F.D.; Souza, R.A.A.; de Farias Neto, S.R.; Lima, A.; et al. Oily Water Separation Process Using Hydrocyclone of Porous Membrane Wall: A Numerical Investigation. *Membranes* **2021**, *11*, 79. [CrossRef]
27. Li, S.; Li, R.; Nicolleau, F.C.; Wang, Z.; Yan, Y.; Xu, Y.; Chen, X. Study on oil–water two-phase flow characteristicsof the hydrocyclone under periodic excitation. *Chem. Eng. Res. Des.* **2020**, *159*, 215–224. [CrossRef]
28. Li, F.; Liu, P.; Yang, X.; Zhang, Y.; Zhao, Y. Effects of inlet concentration on the hydrocyclone separation performance with different inlet velocity. *Powder Technol.* **2020**, *375*, 337–351. [CrossRef]

29. Jing, J.; Zhang, S.; Qin, M.; Luo, J.; Shan, Y.; Cheng, Y.; Tan, J. Numerical simulation study of offshore heavy oil desanding by hydrocyclones. *Sep. Purif. Technol.* **2021**, *258*, 118051. [CrossRef]
30. Lv, W.J.; Dang, Z.H.; He, Y.; Chang, Y.L.; Ma, S.H.; Liu, B.; Gao, L.; Ma, L. UU-type parallel mini-hydrocyclone group for oil-water separation in methanol-to-olefin industrial wastewater. *Chem. Eng. Process. Process Intensif.* **2020**, *149*, 107846. [CrossRef]
31. Liu, L.; Zhao, L.; Yang, X.; Wang, Y.; Xu, B.; Liang, B. Innovative design and study of an oil-water coupling separation magnetic hydrocyclone. *Sep. Purif. Technol.* **2019**, *213*, 389–400. [CrossRef]
32. Yang, M.; Jiang, R.; Wu, X.; Hu, Z.; Yue, Y.; Chen, Y. Numerical analysis of flow field and separation characteristics in an oilfield sewage separation device. *Adv. Powder Technol.* **2021**, *32*, 771–778. [CrossRef]
33. Zeng, X.; Zhao, L.; Zhao, W.; Hou, M.; Zhu, F.; Fan, G.; Yan, C. Experimental Study on a Novel Axial Separator for Oil–Water Separation. *Ind. Eng. Chem. Res.* **2020**, *59*, 21177–21186. [CrossRef]

Article

Growth and Distribution of Coal-Measure Source Rocks in Mixed Platform: A Case Study of Carboniferous in Bamai Area, Southwest Tarim Basin, China

Siyu Su [1,2], Yongqiang Zhao [3], Renhai Pu [1,2,*], Shuo Chen [1,2], Tianyu Ji [4] and Wei Yao [3]

[1] State Key Laboratory of Continental Dynamics, Northwest University, Xi'an 710069, China
[2] Department of Geology, Northwest University, Xi'an 710069, China
[3] SINOPEC Key Laboratory of Petroleum Accumulation Mechanisms, Wuxi Research Institute of Petroleum Geology, Wuxi 214126, China
[4] Research Institute of Petroleum Exploration and Development, PetroChina, Beijing 100083, China
* Correspondence: purenhai@nwu.edu.cn

Citation: Su, S.; Zhao, Y.; Pu, R.; Chen, S.; Ji, T.; Yao, W. Growth and Distribution of Coal-Measure Source Rocks in Mixed Platform: A Case Study of Carboniferous in Bamai Area, Southwest Tarim Basin, China. *Energies* **2022**, *15*, 5712. https://doi.org/10.3390/en15155712

Academic Editors: Gan Feng, Qingxiang Meng, Gan Li and Fei Wu

Received: 14 June 2022
Accepted: 3 August 2022
Published: 5 August 2022

Publisher's Note: MDPI stays neutral with regard to jurisdictional claims in published maps and institutional affiliations.

Abstract: Coal-measure source rocks are generally developed in marsh facies under a humid climate and are rarely reported in a carbonate platform or a mixed platform. Carboniferous seawater intruded from west to east in the Tarim Basin, and mixed platform deposits of interbedded mudstone and carbonate developed in the southwest of the basin. In recent years, with the deepening of the exploration, nearly 20 m coal seams and carbonaceous mudstone source rocks have been found in the Carboniferous lagoon's tidal-flat background. The hydrocarbon generation potential, development, and distribution of these coal-measure source rocks have become an important issue for oil and gas exploration. Coal seams and carbonaceous mudstones were found in the Carboniferous formation of wells BT5, BT10, and Lx2. The hydrocarbon prospect, development, and distribution characteristics of these coal formations have become an important research topic. The authors conducted organic geochemical tests and analyses of core and samples drill cuttings from multiple wells in the study area, combined with research focused on the identification and distribution of coal seams, dark mudstones, and depositional facies via logging cross plots of different lithology and 3D seismic inversion. The results show that coal-measure source rocks in the BT5 well are related to the set of delta-lagoon sedimentary systems widely developed in the Carboniferous Karashayi Formation. The maximum cumulative thickness of coal-measure source rocks is about 20 m, with total organic carbon (TOC) contents of 0.15–60%, kerogen types II_2-III, and vitrinite reflectance (Ro) values of 0.78–1.65%. The rocks have generally low maturity in the northwestern area and high maturity in the southeastern area, and the maturity changes as the burial depth changes. The effective hydrocarbon source rocks such as coal, carbonaceous mudstone, and dark mudstone all show acoustic time (AC) greater than 300 μs/m, and density (DEN) less than 2.3 g/cm^3, but possess different gamma ray (GR) values. The GR value is less than 75 API for coal, between 75–100 API for carbonaceous mudstone, and greater than 100 API for dark mudstones. The distribution of source rocks can be identified in the area between the wells according to a 3D seismic inversion impedance (IMP) of less than 7333 m/s·g/cm^3. The development and controlled factors of coal-measure source rocks of delta facies in the mixed platform have a significant role for oil and gas exploration of Upper Paleozoic in this area. The coal measure and sandstones of delta in the Carboniferous are expected to form self-generation and self-storage pools in this area.

Keywords: Tarim Basin; mixed platform; delta; coal-measure distribution; coal-measure logging response

1. Introduction

The Carboniferous system in the Tarim Basin is generally composed of mixed platform deposits consisting of multiple cycles of carbonate rocks and mudstones (Figure 1c).

From bottom to top, the Carboniferous is composed of 10 stratigraphic members: the Glutenite, Lower Mudstone, Bioclastic Limestone, Middle Mudstone, Standard Limestone of Bachu Formation, Upper Mudstone, Sand–Mudstone, Limestone-Bearing Member of the Karashayi Formation, and Limestone Member of the Xiaohaizi Formation [1,2]. Compared with the central and northern Tarim Basin, the terrigenous clastic proportion in the Carboniferous sediments in the southwest Tarim Basin (SWTB) decreases, and the carbonate content in the mudstone members increases. Furthermore, limestone rocks in the latter atypically contain partial dolomites and rare gypsum [3]. The proportion of carbonate rocks in the Karashayi Formation is a little higher, and coal seams are generally not developed [4,5]. However, multiple layers of coal seams and carbonaceous mudstones occurred in the Karashayi Formation in the BT5 well's seismic survey, suggesting different sedimentary environments and hydrocarbon-generation potential from other areas of the basin. Therefore, it is meaningful to clarify the geochemical features, depositional environment, and distribution of the coal-source rocks.

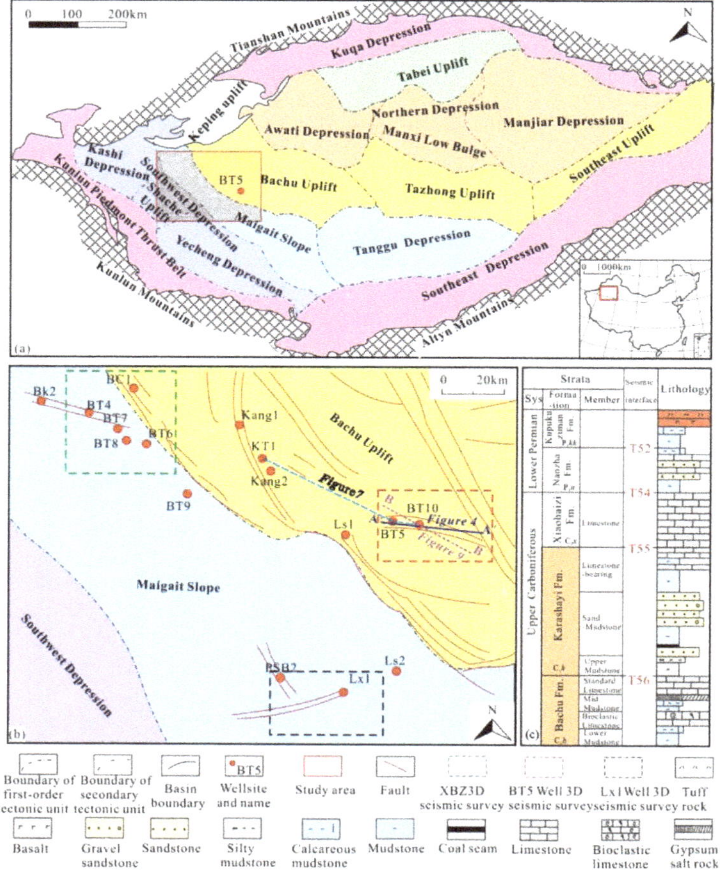

Figure 1. (**a**) The location map of the study area and the division of basin structural units in the Tarim Basin; (**b**) structural units map of the Bamai area, location map of 3D seismic survey, and (**c**) a comprehensive stratigraphic histogram from Carboniferous to Permian in the Bamai area; in (**b**), the green area is the XBZ 3D seismic survey, the red area is the Lx1 well 3D seismic survey, and the black area is the BT5 well 3D seismic survey. In (**c**), the orange area is the target formation.

The Cambrian–Lower Ordovician marine source rocks and the Carboniferous–Lower Permian marine–continental transitional source rocks are present in the Bachu Uplift and Maigaiti slope (abbreviated as the Bamai area) in the SWTB [6–9]. Previous studies have found that the Cambrian–Lower Ordovician source rocks are of high quality [10–13], while research on the Carboniferous source rocks in the Bamai area is relatively rare. Su et al. [14] and Huo et al. [15] conducted oil-source correlation by using carbon isotopes, biomarker compounds, and hydrocarbon components. They concluded that the Bashto light oil mainly formed in the Cambrian, and a small amount may have originated from Carboniferous carbonate and mudstone, as the maturity of the crude oil was obviously higher than that of the Carboniferous source rocks. Shen et al. [16] considered that the two types of source rocks, namely, Carboniferous mixed shelf mudstone and carbonate, contain type II organic matter and are in the overmature evolution stage. Du [17] considered that the Cretaceous oil and gas in well KD1 in the Kunlun Mountains originated from Carboniferous–Permian type II argillaceous mature source rocks in the eastern sag of the SWTB; the TOC contents are 0.04% and 4.19%, respectively, the maximum pyrolysis temperature (Tmax) values are 400–595 °C, and the overall maturity and abundance of organic matter are high. Based on a large number of outcrops and borehole data, Chen et al. [18] mainly studied the lagoon facies argillaceous source rocks of the Karashayi Formation in the Macan 1 well. The source rocks of the Mazhatage structural belt centered on the Macan 1 well were considered to be approximately 60 m, with low to medium maturity and relatively limited distribution.

Based on outcrops and limited drilling data, predecessors mainly studied two types of hydrocarbon source rocks: Carboniferous carbonate rocks and mudstones. Coal-measure source rocks were not been involved, and, furthermore, less attention was paid to the middle and lower mudstones of the Bachu Formation. The methods of logging, seismic attribute [19,20], inversion [21,22], and geochemical analysis are, hereby, used to comprehensively analyze and evaluate the characteristics of hydrocarbon source rocks [23,24]. The well logging and seismic data are used to determine the depositional facies and distribution of hydrocarbon-source rocks [25,26]. This study focuses on the coal seams and carbonaceous mudstones found in the Carboniferous Karashayi Formation in well BT5. The geophysical methods of discriminating different coal-measure source rocks are explored by using well logging and 3D seismic data. At the same time, the seismic facies, depositional environment, and their control over the coals of Karashayi Formation have been analyzed, which provides geological evidences for finding oil and gas reservoirs related to coal-measure source rocks.

2. Geology of the Study Basin

The Tarim Basin is a composite petroliferous basin with an area of approximately 56×10^4 km^2 [27]. The study area is located in two structural units of the Bachu uplift and Maigaiti slope in the southwestern Tarim Basin, China, which is referred to as the Bamai area (Figure 1a). The Bamai area is bordered by the Keping uplift to the north; the northern depression and Tazhong uplift to the east; and the Kashi sag, Shache uplift, and Yecheng sag to the southwest [28–31]. The Carboniferous strata are one of the main subjects of oil and gas exploration in the Bamai area and contain the Bachu Formation, Karashayi Formation, and Xiaohaizi Formation from the bottom up. Under the action of dry and wet climate alternation and the rise and fall of relative sea level, a deep-water shelf, shallow-water shelf, lagoon bay, delta, and other sedimentary environments were mainly developed [32–37]. The main lithologies include limestone, dolomite, marlstone, mudstone, etc., and these lithologies are partially intercalated with gypsum dolomite and gypsum mudstone and locally embedded with coal seams, gypsum dolomite, and gypsum mudstone [38,39]. Years of exploration have shown that the Carboniferous rocks have the capability of hydrocarbon generation, storage and capping [40]. The coal-measure source rocks seen first in the delta environment are related to the Carboniferous mixed platform.

3. Data and Methods

The study area contains 15 exploration wells, 18 2D seismic lines, and 3 3D seismic surveys (Figure 1a), of which the 3D seismic surveys are XBZ 3D seismic surveys (approximately 600 km^2), Lx1 well 3D seismic survey (approximately 350 km^2), and ET5 well 3D seismic survey (approximately 400 km^2). The main frequency of the 3D seismic data is approximately 35 Hz, the sampling rate 2 ms, trace interval 25m, the sampling length 0–6000 ms, and the datum elevation 1000 m. All the seismic data were processed as poststack volumes processed through high resolution, high signal-to-noise ratio, and high relative-amplitude preservation. There are 9 wells in the 3D seismic surveys, most wells drilled through the bottom of Upper Paleozone, with total depth over 4000 m, a few wells have core data of Carboniferous, and each well contains comprehensive logging curves, including gamma ray (GR), resistivity (RD), acoustic time (AC), neutron (CN), density (DEN), etc., with sample rate 0.125 m. Upper Paleozoic revealed by drilling in the area includes Permian Kupukuzman Formation (P$_1$kk), Nanzha Formation (P$_1$n), Xiaohaizi Formation (C$_2$x), Carboniferous Karashayi Formation (C$_1$k), and Bachu Formation (C$_1$b), with bottom seismic reflection interfaces that are T52, T54, T56, and T57, respectively (Figure 1c).

Except for the coring of coal measure in the Karashayi Formation of well BT5, most of the samples of Carboniferous source rocks are drilling cuttings. The coring section of the Karashayi Formation in well BT5 is 2077.5–2084 m, and the lithology from top to bottom is gray black carbonaceous mudstone and dark gray mudstone containing light gray silty strip, with local plant fragments (Figure 2). Coal samples come from drilling cuttings, which are dark black, small, granular, glassy, and stained, and their final affirmatory lithology is determined by subsequent geochemical tests for organic matter.

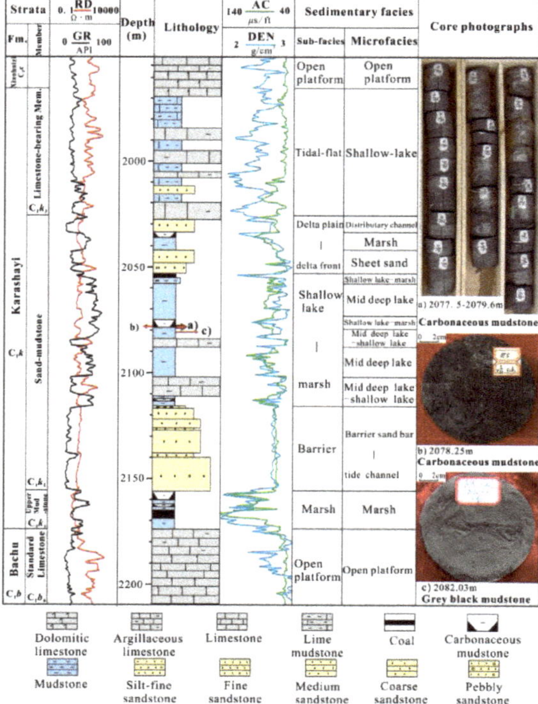

Figure 2. A comprehensive stratigraphic profile of the Carboniferous Karashayi Formation of well BT5.

Based on the core, logging, and cutting data of BT5 and BT10 wells in the work area, cross plots of AC-GR and DEN-RD to differentiate lithologies are established. The threshold values for quantitative identification of coal seams, carbonaceous mudstone, shale, sandstone, and carbonate rocks are determined. Then, synthetic seismograms of two wells, BT5 and BT10, are made, well seismic calibration is completed, and log-constrained poststack wave impedance inversion is carried out in the 3D seismic survey of BT5. In the invertomer, the content and thickness percentages of the related lithologies of the Karashayi Formation are extracted, according to the wave impedance of the coal seam, carbonaceous mudstone, and sandstone.

In addition, 30 samples of Carboniferous cores and drilling cuttings are collected from Sinopec; in particular, these samples include coal, carbonaceous mudstone, and dark mudstone from wells BT5, BT10, and BT7. After sample separation and selection of different lithologies, 12 samples of drill cuttings and 2 core samples from the Carboniferous Karashayi and Bachu Formations are obtained. The organic matter extraction and pyrolysis experiments of all 14 samples are carried out in the State Key Laboratory of Continental Dynamics of Northwestern University, and the organic carbon content, type, and maturity of each sample are determined (Table 1).

Table 1. TOC, Ro, and organic carbon type of hydrocarbon source rock samples from the Carboniferous Karashayi Formation and Bachu Formation.

Number	Well	Stratum	Sample Type	Depth (m)	Lithology	T_{max} (°C)	TOC (%)	Ro	Organic Carbon Type
1	BT5	C_1k_2	Core	1918.1	Carbonaceous mudstone	437	14.2	1.65	III
2	BT5	C_1k_2	Core	1917.6	Mudstone	434	0.15	0.90	III
3	KT1	C_1k_2	Cuttings	1620	Carbonaceous mudstone	431	14.4	0.78	II_2
4	BT5	C_1k_2	Cuttings	2040	Carbonaceous mudstone	424	10.6	0.78	II_2-III
5	BT10	C_1b_1	Cuttings	2660	Mudstone	436	0.01	0.80	III
6	BT7	C_1k_2	Cuttings	4520	Mudstone	431	0.01	0.97	III
7	BT7	C_1k_2	Cuttings	4540	Mudstone	428	0.86	1.33	III
8	BT9	C_1k_2	Cuttings	4740	Mudstone	437	1.15	1.13	III
9	BT5	C_1k_1	Cuttings	2020	Mudstone	428	2.36	0.87	III
10	BT5	C_1k_2	Cuttings	2095	Mudstone	432	0.94	0.80	III
11	BT5	C_1b_3	Cuttings	2400	Mudstone	429	0.15	/	III
12	BT5	C_1k_2	Cuttings	2055	Coal	425	58.1	1.29	II_1-II_2
13	BT5	C_1k_2	Cuttings	2060	Carbonaceous mudstone	424	18.5	0.81	II_2
14	KT1	C_1k_2	Cuttings	1605	Coal	432	63.2	0.93	III

By plotting the Ro value and well depth of source rock samples measured in different wells, the trend of source rock maturity with target layer depth in the study area is obtained. The trend is then extrapolated to the structural map of the bottom of the Karashayi Formation to estimate the maturity of the source rocks of the said formation in the Bamai area.

4. Results

4.1. Logging Identification of Carboniferous Source Rocks

Logging curves calibrated by the core (Figure 2) and AC-GR cross plot (Figure 3) indicate that coal and dark mudstone have logging responses of low DEN, high AC, and low RD values [41], and their AC values are all more than 300 μs/m, the DEN value is generally less than 2.2 g/cm^3, and the RD value is less than 50 Ω·m; they are easy to be identified in logging curves because of their high-amplitude spikes. Coal, carbonaceous mudstone, and dark shale can be distinguished by GR values. The GR values are <75 API for coal seam, between 75 and 100 API for carbonaceous mudstone, and >100 API for dark mudstone. According to the above logging identification method, two coal seam layers are present in the BT5 well, with a cumulative thickness of approximately 6 m; five layers of carbonaceous mudstone are present with a cumulative thickness of approximately 10 m; and nine layers of dark mudstone are present with a total thickness of approximately 66 m, of which coal seams and carbonaceous mudstone account for 24.2%. The sandstone shows a box-shaped GR logging curves [42], with AC < 240 μs/m and GR < 60 API, and RD < 100 Ω·m, which is much less than the RD values of carbonate rocks (Figure 3).

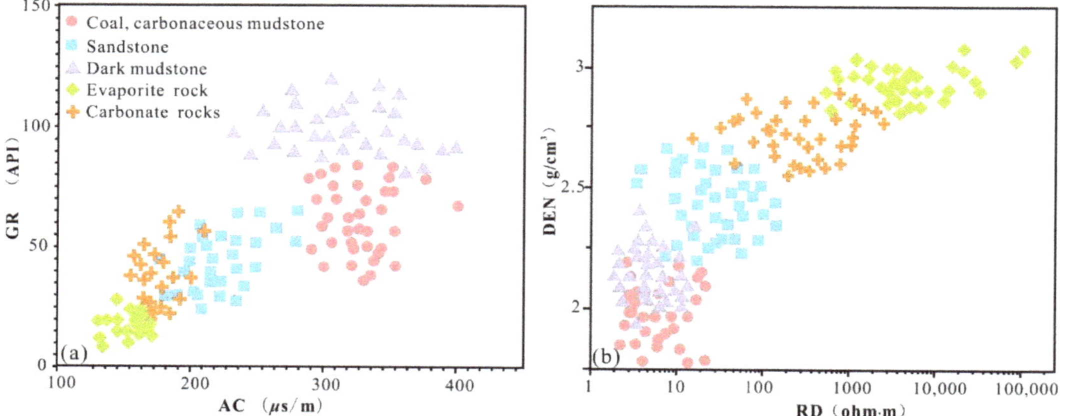

Figure 3. Log cross plots of different lithologies of the Karashayi Formation in the Bamai area. (**a**) AC-GR cross plot. (**b**) DEN-RD cross plot.

4.2. 3D Seismic Identification and Distribution of Coal-Measure Source Rocks in the Karashayi Formation

In order to identify coal-measure source rocks with 3D seismic data, we have performed log constrained 3D seismic impedance inversion. Figure 4 is the BT5–BT10 well-linked inversion acoustic section. There is a bright yellow banded layer between seismic interfaces from T52 to T54 that corresponds to a large set of low-speed mudstone and tuff widely distributed in the lower Permian. The seismic interface from T55 to T56 corresponding to the Carboniferous Karashayi Formation, with an AC value of 200–350 μs/m, about 10–40 ms thick, in which the part of AC > 300 μs/m should belong to coal-measure source rocks, and they gradually thicken to the southeast lower part of the depression.

According to the logging and seismic-identification threshold values of coal seam and carbonaceous mudstone of the Karashayi Formation in the study area (the AC > 300 μs/m, DEN < 2.2 g/cm^3, and Imp < 7333 m/s·g/cm^3), the percentage content of wave impedance less than 7333 m/s·g/cm^3 in the Karashayi Formation can be extracted from the 3D seismic-wave impedance inversion volume as the percentage map of coal-measure source rocks (Figure 5). The map indicates that the coal and carbonaceous mudstone are thickening to the southeast.

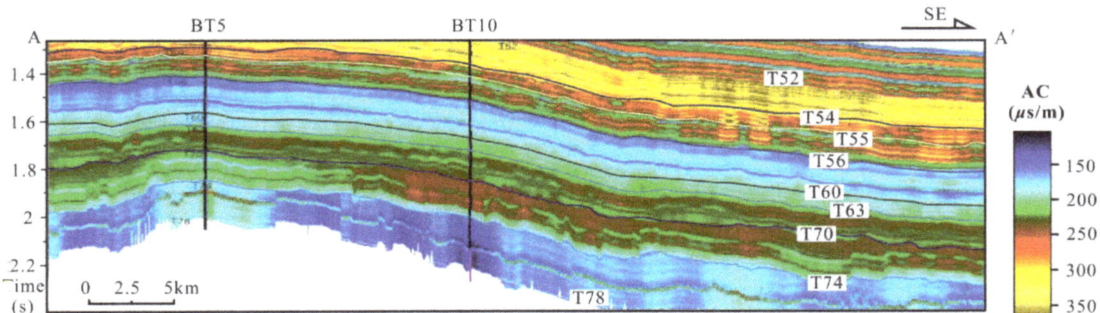

Figure 4. BT5–BT10 well-linked 3D seismic inversion acoustic profile. See Figure 1b for section position.

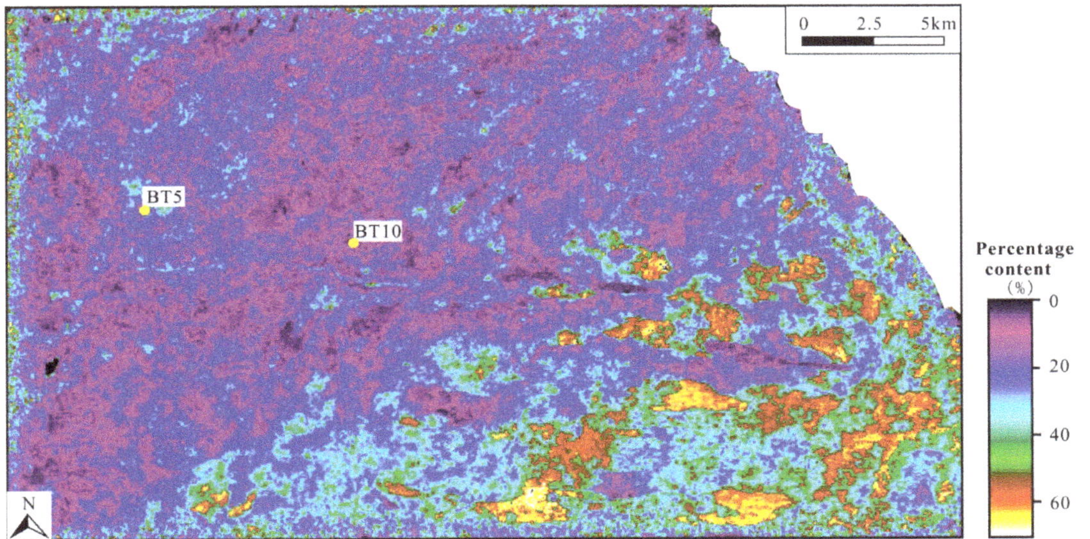

Figure 5. Percentage contents of coal and carbonaceous mudstone in the Karashayi Formation in the BT5 well 3D seismic survey.

Using the SUM module of landmark, the cumulative thickness of source rocks in the Karashayi Formation of the 13 drilling well points is counted, and, under the constraint of the percentage of coal seams, the carbonaceous mudstone and dark mudstone are obtained through 3D seismic inversion, so the contour map of the thicknesses of coal seams and carbonaceous mudstones in the study area is drawn (Figure 6). The map shows that the thicknesses of the coal-measure source rock of the Karashayi Formation in the study area are about 0–20 m and thicken from northwest to southeast. The area of thick coal-measure source rocks is about 40 km^2.

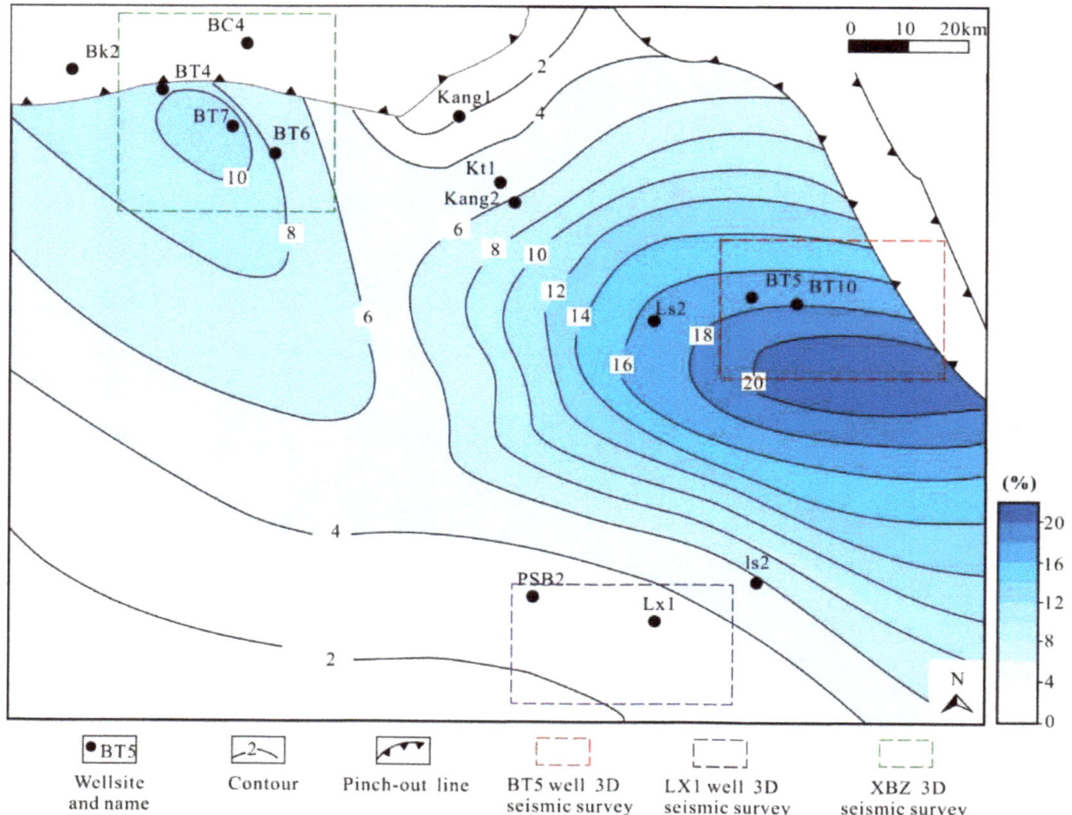

Figure 6. Isopach map of the cumulative thickness of the Carboniferous Karashayi Formation coal and carbonaceous mudstone in the Bamai area.

4.3. Geochemical Characteristics of Source Rocks

The test results of samples from wells BT5, BT10, KT1, BT7, and BT9 show that the TOC contents of the Karashayi Formation coal range from 58.1% to 63.2%, with an average of 60.65%, and the Ro values range from 0.93% to 1.29%, with an average of 1.11%. The TOC contents of carbonaceous mudstone range from 10.6% to 18.5%, with an average of 14.43%, and the Ro values range from 0.78% to 1.65%, with an average of 1.00%. The TOC contents of dark mudstone range from 0.01% to 2.36%, with an average of 0.91%, and the Ro values range from 0.80% to 1.33%, with an average of 1.00%. According to the calculation results of the hydrogen index–Tmax diagram [43], the organic carbon types of most coal seams and carbonaceous mudstones are of type III, and some carbonaceous mudstones are of types II$_2$-III (Table 1).

Based on the test results of the Carboniferous Karashayi Formation coal and dark mudstone, the TOC content is between 0.17% and 63.2%, the types of organic carbon are II$_2$-III, and the Ro values are 0.78–1.65% for coal and dark mudstone of the Carboniferous Karashayi Formation. The Tmax values range from 424 °C to 437 °C, with an average of 430 °C, indicating oil generation of immature to low maturity, which are slightly lower than the maturity indicated by the Ro values.

The preliminary sample separation and selection revealed that most of the samples collected from the Middle and Lower Mudstone Members of the Carboniferous Bachu Formation do not meet the requirements of pure dark mudstone, are not distinguishable from dry drilling mud, or are oxidized brown. Only two samples from the Middle and Lower Mudstone Members meets the test conditions (see Yellow Sampling Points in Figure 7). Test shows that the TOC contents of the dark shale in the Middle and Lower Mudstone Members of the Carboniferous Bachu Formation are low, ranging from 0.01–0.15%, with an average of 0.08%, which does not meet the conditions for hydrocarbon generation (Table 1).

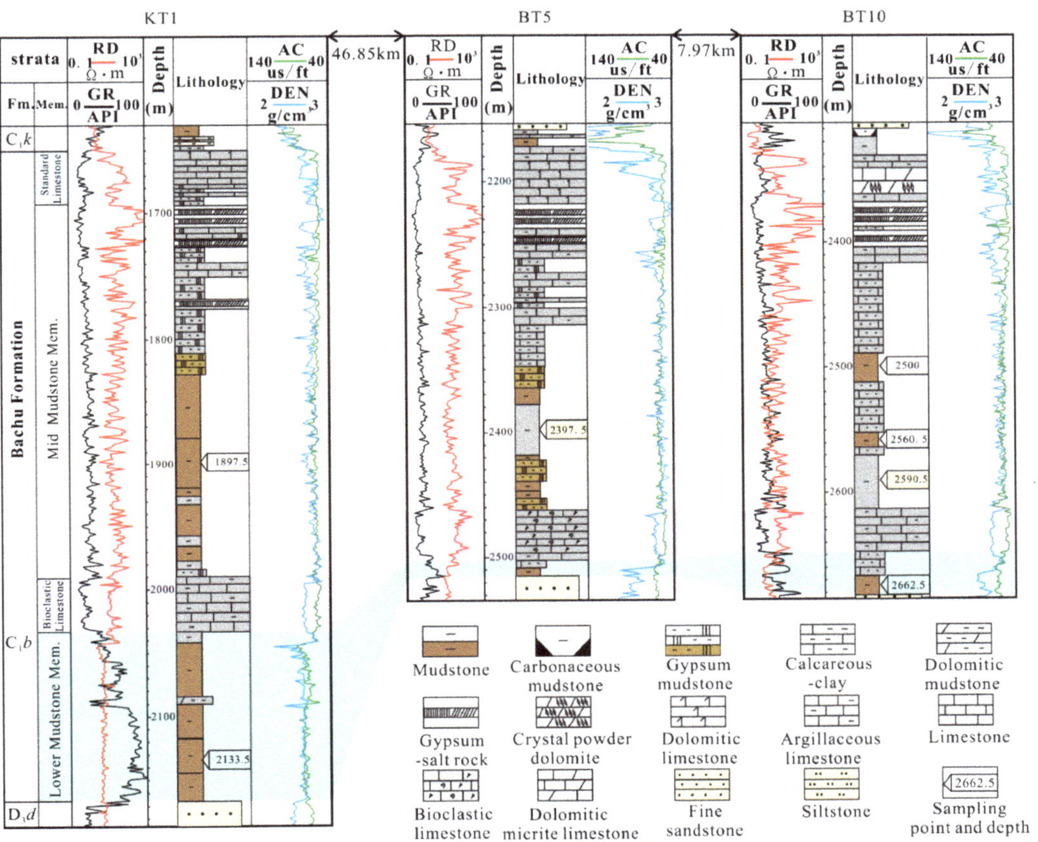

Figure 7. Profile of wells KT1, BT5, and BT10 of the Bachu Formation. See Figure 1b for the lateral positions. In the legend, the different colors filled by mudstone and gypsum mudstone reflect different sedimentary backgrounds, in which brown represents the oxidation environment and gray represents the reduction environment.

In Table 1, the linear relationship of Ro = 0.0001H + 0.622 between the Ro and sample depth (H) is obtained by linear fitting. The depth map of the bottom boundary of the Karashayi Formation in the Bamai area can be made through the two-way time map of the seismic interface T56 and the time–depth conversion of the Karashayi Formation. By using the linear relationship of Ro–H, the depth map of the top surface of the Karashaiyi Formation can be transformed into a contour map showing vitrinite reflectance (Figure 8). The map shows that the Ro value of the source rock of the Karashayi Formation in the Bamai area is between 0.7% and 1.3%, which has a tendency to increase toward the southwest.

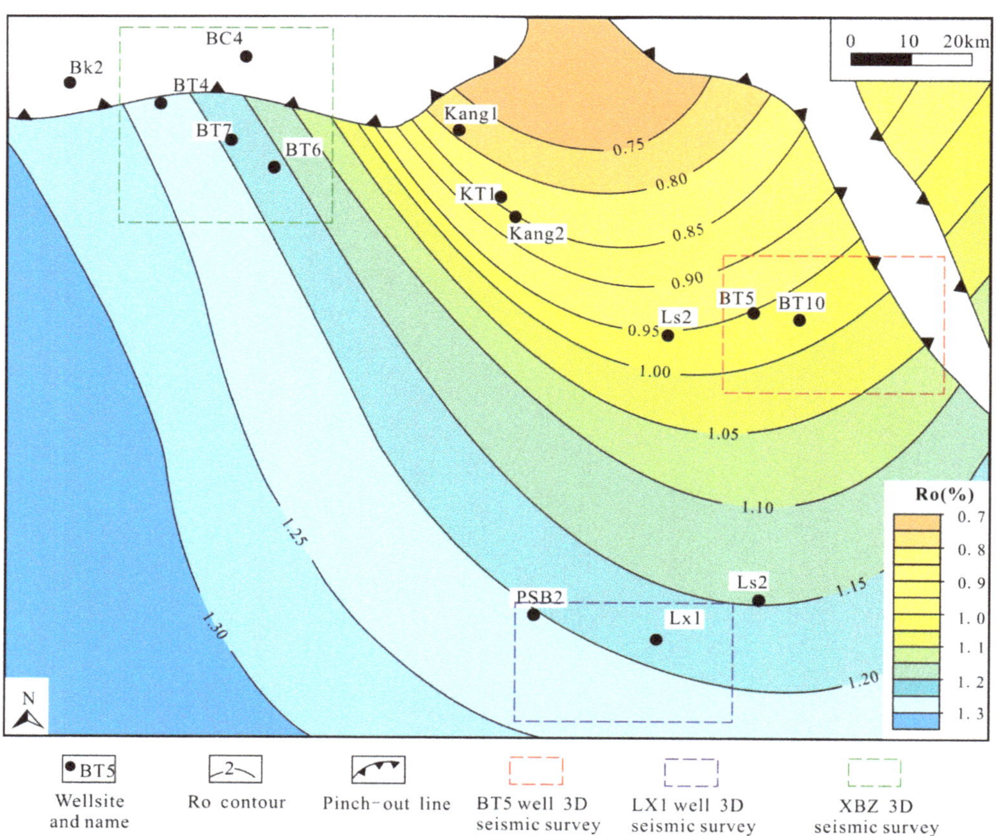

Figure 8. Vitrinite reflectance Ro contour map of the bottom Carboniferous Karashayi Formation in the Bamai area.

4.4. Deltaic Control on Coal-Measure Source Rocks

A set of imbricated progradational reflection can be identified in the Carboniferous Karashayi Formation of the BT5 3D seismic survey, which advances from northwest to southeast. The set is prograded by the phase axes of five progradational wave peaks with medium and strong amplitudes, representing the five stages of progradational sedimentary bodies (Figure 9). The progradational reflections can exist in a variety of sedimentary environments [44]. The seismic reflections of the five-stage progradation are traced, interpreted, and closed in detail in the BT5 well 3D seismic survey, and the plane distribution of the five stage progradation is obtained (Figure 10). BT5 and BT10 are located on the first and the second progradational leaves, respectively, and they correspond to the delta-front facies, with the upward coarsening of the Karashayi Formation (Figure 2). Based on the regional drilling data and previous studies [45], this delta system was developed in the lagoon sedimentary background and surrounded by a tidal flat and barrier system.

Sandstone content has a relatively sensitive response to sedimentary microfacies, and is an important means to study depositional facies and sand-body distribution rules [23,24]. Different sandstone content can reflect different sedimentary environments. Based on the data of 13 wells situated in the Bamai area, the percentage content of sandstone in the Karashayi Formation was estimated, and its depositional facies map was made accordingly (Figure 11). Due to the limited number of wells in the study area, the sandstone content is linearly extrapolated in the non-drilled area (such as the southwest area), so the depositional facies are not described in detail, but the genesis, distribution, and identification of coal-measure source rocks are focused in the general context of clarifying that the depositional facies is a mixed platform-lagoon tidal-flat environment. The study area is a lagoonal sedimentary setting overall, in which three delta front lobes from the northwest are developed, and the percentage of sandstone in the delta is high, ranging from 25% to 45%. The depositional facies in the western part of the study area are transitional to the mixed platform and the tidal-flat subfacies.

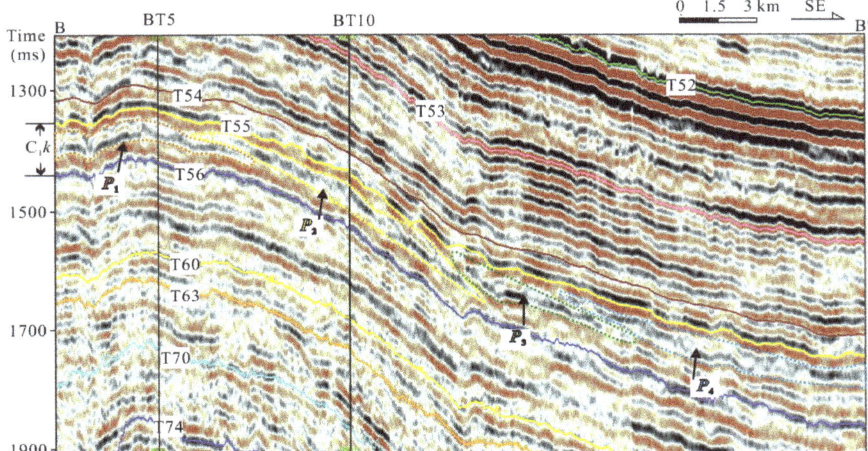

Figure 9. Seismic section across wells BT5 and BT10 in SE direction of the delta progradational reflection of the Karashayi Formation. See Figure 1b for section position, P_1 means progradation I, P_2 means progradation II, P_3 means progradation III, and P_4 means progradation IV. Karashayi Formation is located between T55–T56 seismic reflection boundaries. See Figure 1b for the location.

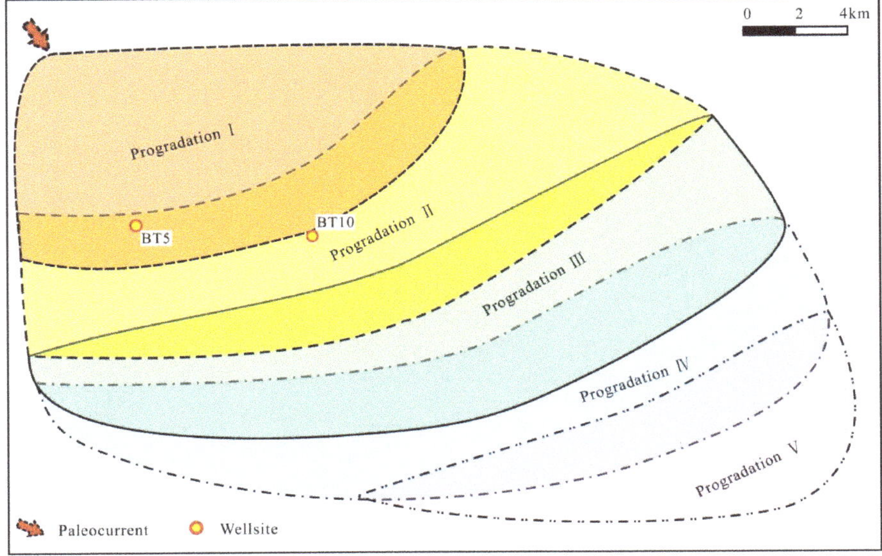

Figure 10. Superposition map five-stage progradational reflections in five different semitransparent colors.

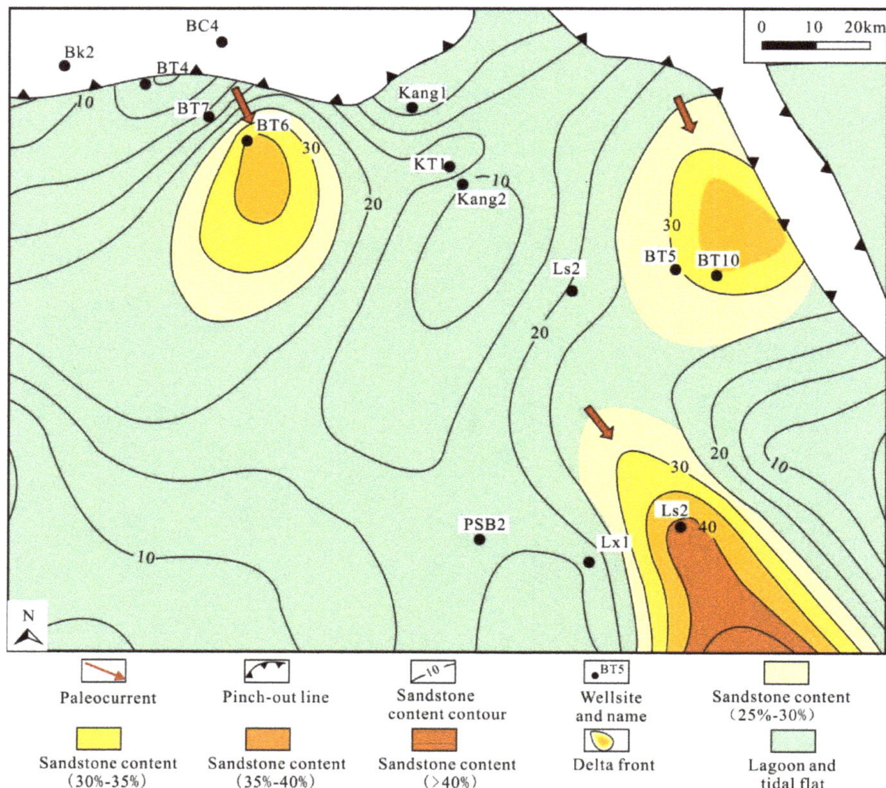

Paleocurrent
Pinch-out line
Sandstone content contour
Wellsite and name
Sandstone content (25%–30%)
Sandstone content (30%–35%)
Sandstone content (35%–40%)
Sandstone content (>40%)
Delta front
Lagoon and tidal flat

Figure 11. Depositional facies map of the Carboniferous Karashayi Formation in the Bamai area.

5. Discussion

5.1. Relationship between the Coal-Measure Source Rocks and Deltas in the Background of Mixed Platform

A comparison of Figures 7 and 11 shows that the thicker area of the coal-measure source rocks in the Bamai area is basically consistent with the distribution area of the delta front; this result indicates that in the lagoonal sedimentary background of the Karashayi Formation as a whole in the southwestern Tarim Basin area, there was an inflow of a freshwater river from the northwest in the BT5 well 3D seismic survey, and different progradational deltas were accumulated gradually from northwest to southeast. This process formed not only sandstone deposits, such as an estuary bar and distributary channels in the delta front, but also a certain thickness of swamp coal and carbonaceous mudstone on the delta plain. Since the organic matter contents of coal and carbonaceous mudstones are much higher than those of lagoon facies' dark mudstones, the coal-measure source rocks in the delta sedimentary area should have better hydrocarbon-generation potential than other surrounding areas.

Coal seams are commonly found in clastic strata and deposited in swampy environments in humid climates [46]. The Carboniferous in the Tarim Basin contains interbedded arid and semi-arid carbonate rocks and mudstones, with local gypsum salts, and no coal developed in other parts of the basin or the surrounding outcrops. Therefore, the Karashayi Formation coal seams in the BT5 well 3D seismic area are a special case, and their formation and distribution should be related to the freshwater delta that occurred locally in the mixed platform. Carbonate rocks are deposited in clear, warm, and full sunlight in a shallow water environment, while mudstone and gypsum mudstone are deposited in hot and turbid shallow water or a tidal-flat environment [47]. The local existence of a delta makes the water bodies less saline, becoming turbid, which weakens the sedimentation of the carbonate rock and evaporated salt rock and increases the proportion of local sand–mud sediment in the background of the mixed platform. In addition, the fresh water body of the delta plain is beneficial to the growth of vegetation to form coal measures. The upward coarsening cycles of Carboniferous Karashayi Formation in the BT5 well and the typical imbricate progradational reflections in the 3D seismic survey verify the existence of deltas. The low seismic-inversion impedance shows that coal seams are also thickened in the delta sedimentary area, further suggesting the close relationship between the origin and distribution of coal measures and the delta in this area.

5.2. Potential of the Coal-Measure Source Rocks in Bamai Area

Many years of exploration practice and previous studies have shown that the Cambrian–Lower Ordovician source rocks provide oil and gas for the Ordovician gas fields such as Yubei, Hetian River, Niaoshan, BT5 well, etc., in the SWTB [32]. The Carboniferous source rocks in the Bamai area have a certain role in hydrocarbon generation and may be used as the minor source rocks of the Bashito and Yasundi oil and gas reservoirs. Two types of hydrocarbon source rocks were considered to exist in the Carboniferous: shallow-sea mixed continental-shelf mudstone and carbonate rock with low maturity, which only reached the maturity stage near well Kedong 1 in Kunlun Mountains, while coal-measure source rocks were never found before [17]. Our study reveal that there is a set of good source rocks in the Bamai area, with the maturity increasing with burial depth, and the content of the organic carbon is much higher than that of the dark mudstone in the same stratum. Coal measures are of better hydrocarbon generation potential and, hopefully, are one of the main source rocks in this area like Cambrian shale.

5.3. Excellent Conditions for Self-Generation and Self-Storage Composite Reservoirs in Carboniferous

In the Bamai area, the oil and gas pools related to the Cambrian–Ordovician source rocks are distributed along the faults, which is a necessary passage for oil and gas migration vertically upward [12]. However, in the areas where Carboniferous high-quality coal-measure source rocks exist, reservoirs could be formed without fault passages. The high-quality coal source rocks of Carboniferous Karashayi Formation in Bamai area have a substantial thickness and wider distribution. They are vertically superposed and transversely contiguous to delta-front sandstones. The deltaic sand body is pinched out toward the upper northwest, satisfying the conditions to form stratigraphic traps. The oil and gas generated by coal-measure source rocks and delta sandstones constitute a beneficial source-reservoir-cap matching, trap, and migration condition. Therefore, the "self-generating, self-accumulating" reservoirs of the Karashayi Formation are expected to be an important exploration target in this area.

6. Conclusions

1. A set of coal and carbonaceous mudstone source rocks are revealed in the Karashayi Formation in the SWTB, China, with an average TOC content of 60.6% and 14.4%, respectively, for coal and carbonaceous mudstone, with types II_2-III kerogen, and Ro values of 0.78–1.65%.

2. The coal seam and carbonaceous mudstone of the Karashayi Formation are characterized by low DEN (<2.2 g/cm^3), high AC (>300 μs/m), and low wave impedances (<7333 m/s·g/cm^3). They can be roughly distinguished according to the natural gamma logging values less and greater than 75 API, respectively. The thicknesses of a single-layer coal seam drilled by wells are generally 1–4 m, the cumulative thickness is approximately 20 m plus carbonaceous mudstones, and the area of thick coal-measure source rocks is about 40 km^2.

3. The Karashayi Formation in the Bamai area is mainly a delta-lagoon deposit in a mixed-platform background. Coal and carbonaceous mudstone are mainly distributed along the delta system and thin toward the tidal-flat area in the west. The hydrocarbon potential of the coal-measure source rocks in the delta of the Karashayi Formation near well BT5 should be better than that in the adjacent area overall.

4. The Lower Mudstone Member and the Middle Mudstone Member of the Bachu Formation contain interbedded gray and variegated mudstones. The TOC contents of dark mudstones with single-layer thicknesses of more than 5 m are generally less than 0.3%, and the mudstones do not have the conditions for hydrocarbon generation.

Author Contributions: Writing—original draft, S.S., Y.Z., R.P., S.C., T.J. and W.Y. Conceptualization, R.P. and S.S.; methodology, S.S.; software, S.C.; validation, Y.Z., R.P. and S.S.; formal analysis, R.P.; investigation, S.C.; resources, Y.Z.; data curation, S.S.; writing—original draft preparation, S.S.; writing—review and editing, R.P.; visualization, S.S.; supervision, W.Y.; project administration, R.P.; funding acquisition, T.J. All authors have read and agreed to the published version of the manuscript.

Funding: This research was funded by "Reservoir forming conditions and zonal evaluation of Paleozoic Clastic Rocks in Tarim" grant number [No. p21049-2].

Acknowledgments: The authors thank the Sinopec Northwest Oilfield Branch for providing relevant 3D seismic and well data.

References

1. Ren, Q.Q.; Feng, J.W.; Johnston, S.; Zhao, L.B. The influence of argillaceous content in carbonate rocks on the 3D modeling and characterization of tectonic fracture parameters—example from the carboniferous and ordovician formations in the hetianhe gas field, Tarim Basin, NW China. *J. Pet. Sci. Eng.* **2021**, *203*, 108668. [CrossRef]

2. Fu, M.Y.; Song, R.C.; Gluyas, J.; Zhang, S.N.; Huang, Q. Diagenesis and reservoir quality of carbonates rocks and mixed siliciclastic as response of the Late Carboniferous glacio-eustatic fluctuation: A case study of Xiaohaizi Formation in western Tarim Basin. *J. Pet. Sci. Eng.* **2019**, *177*, 1024–1041. [CrossRef]

3. Zhu, G.Y.; Zhang, S.C.; Su, J.; Meng, S.C.; Yang, H.J.; Hu, J.F.; Zhu, Y.F. Secondary accumulation of hydrocarbons in Carboniferous reservoirs in the northern Tarim Basin, China. *J. Pet. Sci. Eng.* **2013**, *102*, 10–26. [CrossRef]

4. Geng, F.; Wang, H.X.; Hao, J.L.; Gao, P.B. Internal Structure Characteristics and Formation Mechanism of Reverse Fault in the Carbonate Rock, A Case Study of Outcrops in Xike'er Area, Tarim Basin, Northwest China. *Front. Earth Sci.* **2021**, *9*, 1165. [CrossRef]

5. Wu, X.Q.; Tao, X.W.; Hu, G.Y. Geochemical characteristics and source of natural gases from Southwest Depression of the Tarim Basin, NW China. *Org. Geochem.* **2014**, *74*, 106–115. [CrossRef]

6. Bian, L.H.; Liu, X.F.; Wu, N.; Wang, J.; Liang, Y.L.; Ni, X.E. Identification of High-Quality Transverse Transport Layer Based on Adobe Photoshop Quantification (PSQ) of Reservoir Bitumen: A Case Study of the Lower Cambrian in Bachu-Keping Area, Tarim Basin, China. *Energies* **2022**, *15*, 2991. [CrossRef]

7. Pang, X.Q.; Chen, J.Q.; Li, S.M.; Chen, J.F. Evaluation method and application of the relative contribution of marine hydrocarbon source rocks in the Tarim basin: A case study from the Tazhong area. *Mar. Petro. Geol.* **2016**, *77*, 1–18. [CrossRef]

8. Song, Z.H.; Tang, L.J.; Liu, C. Variations of thick-skinned deformation along tumuxiuke thrust in Bachu uplift of Tarim Basin, northwestern China. *J. Struct. Geol.* **2021**, *144*, 104277. [CrossRef]

9. Cui, H.F.; Tian, L.; Liu, J.; Zhang, N.C. Hydrocarbon discovery in the Ordovician dolomite reservoirs in the Maigaiti slope, Southwest Depression of the Tarim Basin, and its exploration implications. *NGIB* **2017**, *4*, 354–363. [CrossRef]

10. Chang, J.; Yang, X.; Qiu, N.S.; Min, K.; Li, C.X.; Li, H.L.; Li, D. Zircon (U-Th)/He thermochronology and thermal evolution of the Tarim Basin, Western China. *J. Asian Earth Sci.* **2022**, *230*, 105210. [CrossRef]

11. Xiao, Z.L.; Hao, Q.Q.; Shen, Z.M. Maturity Evolution of the Cambrian Source Rocks in the Tarim Basin. *Adv. Mat. Res.* **2013**, *622–623*, 1642–1645. [CrossRef]

12. Hu, G.; Meng, Q.Q.; Wang, J.; Tengger; Xie, X.M.; Lu, L.F.; Luo, H.Y.; Liu, W.H. The Original Organism Assemblages and Kerogen Carbon Isotopic Compositions of the Early Paleozoic Source Rocks in the Tarim Basin, China. *Wiley-Blackwell* **2018**, *92*, 2297–2300. [CrossRef]

13. Wu, J.Y.; He, X.Y.; Ni, X.F.; Huang, L.L. Characteristics of the Lower Cambrian Yuertus Formation source rocks in the Tarim Basin, NW China. *EES* **2019**, *360*, 12–43. [CrossRef]

14. Su, L.; Zhang, D.W.; Zheng, L.J.; Liu, X.W.; Yang, X.; Zheng, J.J.; Liu, P. Experimental study of the influences of pressure on generation and expulsion of hydrocarbons: A case study from mudstone source rocks and its geological application in the Tarim Basin. *J. Pet. Sci. Eng.* **2020**, *189*, 107021. [CrossRef]

15. Huo, Z.; Pang, X.; Ouyang, X.; Zhang, B.; Shen, W.; Guo, F.; Wang, W. Upper limit of maturity for hydrocarbon generation in carbonate source rocks in the Tarim Basin Platform, China. *Arabian J. Geosci.* **2015**, *8*, 2497–2514. [CrossRef]
16. Shen, W.B.; Pang, X.Q.; Jiang, F.J.; Zhang, B.S.; Shen, W.B.; Guo, F.T.; Wang, W.Y. Accumulation model based on factors controlling Ordovician hydrocarbons generation, migration, and enrichment in the Tazhong area, Tarim Basin, NW China. *Arabian J. Geosci.* **2016**, *9*, 347. [CrossRef]
17. Du, Z.L.; Zeng, C.M.; Qiu, H.J.; Yang, Y.X.; Zhang, L. Characteristics of two sets of Permian source rocks and oil source analysis of well Kedong 1 in Yecheng sag, southwestern Tarim Basin. *J. Jilin Univ.* **2016**, *46*, 651–660. [CrossRef]
18. Chen, J.Q.; Pang, X.Q.; Pang, H. Quantitative prediction model of present-day low-TOC carbonate source rocks: Example from the Middle-upper Ordovician in the Tarim Basin. *Energ. Explor. Exploit.* **2018**, *36*, 1335–1355. [CrossRef]
19. Anees, A.; Shi, W.Z.; Ashraf, U.; Xu, Q.H. Channel identification using 3D seismic attributes in lower Shihezi Formation of Hangjinqi area, northern Ordos Basin, China. *J. Appl. Geophys.* **2019**, *163*, 139–150. [CrossRef]
20. Ashraf, U.; Zhang, H.C.; Anees, A.; Mangi, H.N.; Tan, S.C. A core logging, machine learning and geostatistical modeling interactive approach for subsurface imaging of lenticular geobodies in a clastic depositional system, se pakistan. *Nat. Resour. Res.* **2021**, *30*, 2807–2830. [CrossRef]
21. Haitham, H.; Adam, P.; Larry, L. Prestack structurally constrained impedance inversion. *Geophysics* **2018**, *83*, 89–103. [CrossRef]
22. Ashraf, U.; Zhang, H.C.; Anees, A.; Ali, M.; Zhang, X.N.; Abbasi, S.S.; Mangi, H.N. Controls on Reservoir Heterogeneity of a Shallow-Marine Reservoir in Sawan Gas Field, SE Pakistan: Implications for Reservoir Quality Prediction Using Acoustic Impedance Inversion. *Water* **2020**, *12*, 2972. [CrossRef]
23. Anees, A.; Zhang, H.C.; Ashraf, U.; Wang, R.; Liu, K.; Mangi, H.N.; Jiang, R.; Zhang, X.N.; Liu, Q.; Tan, S.C.; et al. Identification of Favorable Zones of Gas Accumulation via Fault Distribution and Depositional facies: Insights from Hangjinqi Area, Northern Ordos Basin. *Front. Earth Sci.* **2022**, *9*, 1375. [CrossRef]
24. Anees, A.; Zhang, H.C.; Ashraf, U.; Wang, R.; Liu, K.; Abbas, A.; Ullah, Z.; Zhang, X.N.; Duan, L.Z.; Liu, F.W.; et al. Depositional facies Controls for Reservoir Quality Prediction of Lower Shihezi Member-1 of the Hangjinqi Area, Ordos Basin. *Minerals* **2022**, *12*, 126. [CrossRef]
25. Michael, P.M.; Robert, L.N.; Anne, M.T. Seismic attribute inversion for velocity and attenuation structure using data from the GLIMPCE Lake Superior experiment. *JGR* **1997**, *102*, 9949–9960. [CrossRef]
26. Thanh, H.V.; Sugai, Y.C. Integrated modelling framework for enhancement history matching in fluvial channel sandstone reservoirs. *Upstream Oil Gas Technol.* **2021**, *6*, 100027. [CrossRef]
27. Li, Y.J.; Wen, L.; Yang, H.J.; Zhang, G.Y.; Shi, J.; Peng, G.X.; Hu, J.F.; Luo, J.C.; Huang, Z.B.; Chen, Y.G. New discovery and geological significance of Late Silurian–Carboniferous extensional structures in Tarim Basin. *J. Asian Earth Sci.* **2015**, *98*, 304–319. [CrossRef]
28. Yu, J.B.; Zhang, J.; Shi, B.P. A Study on Tectono-Thermal Evolution History of Bachu Uplift in the Tarim Basin. *Chin. J. Geophys.* **2010**, *53*, 805–814. [CrossRef]
29. Zhou, X.B.; Li, J.H.; Li, W.S.; Cheng, Y.L.; Wang, H.H. Tectonic Units Division of Crystalline Basement in the Tarim Basin, NW China. *Acta Geol. Sin.* **2015**, *89*, 127. [CrossRef]
30. Neng, Y.; Wu, G.H.; Huang, S.Y.; Zhang, X.; Cao, S.J. Formation stage and controlling factors of the paleo-uplifts in the Tarim Basin: A further discussion. *NGIB* **2016**, *3*, 209–215. [CrossRef]
31. Wang, Z.Y.; Gao, Z.Q.; Fan, T.L.; Zhang, H.H.; Yuan, Y.X.; Wei, D.; Qi, L.X.; Yun, L.; Karubandika, G.M. Architecture of strike-slip fault zones in the central Tarim Basin and implications for their control on petroleum systems. *J. Pet. Sci. Eng.* **2022**, *213*, 110432–114032. [CrossRef]
32. Zhang, J.L.; Zhang, Z.J. Depositional facies of the Silurian tide-dominated paleo-estuary of the Tazhong area in the Tarim Basin. *Pet. Sci.* **2008**, *5*, 95–104. [CrossRef]
33. Lin, C.S.; Yang, H.J.; Liu, J.Y.; Peng, L.; Cai, Z.Z.; Yang, X.F.; Yang, Y.H. Paleostructural geomorphology of the Paleozoic central uplift belt and its constraint on the development of depositional facies in the Tarim Basin. *Sci. China Earth Sci.* **2009**, *52*, 823–834. [CrossRef]
34. Liu, K.N.; Jiang, Z.X.; Zhong, D.K.; Kang, R.; Cao, Y. Study on Depositional facies of Carboniferous Kalashayi Formation in the South of Tahe Oilfield. *Adv. Mat. Res.* **2012**, *433*, 1745–1750. [CrossRef]
35. Ma, B.B.; Lu, Y.C.; Eriksson, K.A.; Peng, L.; Xing, F.C.; Li, X.Q. Multiple organic–inorganic interactions and influences on heterogeneous carbonate-cementation patterns: Example from Silurian deeply buried sandstones, central Tarim Basin, northwestern China. *Sedimentology* **2020**, *68*, 670–696. [CrossRef]
36. Kassem, A.A.; Hussein, W.S.; Radwan, A.E.; Anani, N.; Abioui, M.; Jain, S.; Shehata, A.A. Petrographic and diagenetic study of siliciclastic Jurassic sediments from the northeastern margin of Africa: Implication for reservoir quality. *J. Petrol. Sci. Eng.* **2021**, *200*, 108340. [CrossRef]
37. Kassem, A.A.; Osman, O.; Nabawy, B.; Baghdady, A.R.; Shehata, A.A. Microfacies Analysis and Reservoir Discrimination of Channelized Carbonate Platform Systems: An Example from the Turonian Wata Formation, Gulf of Suez, Egypt. *J. Petrol. Sci. Eng.* **2022**, *212*, 110272. [CrossRef]
38. Liu, L.F.; Qiao, J.Q.; Shen, B.J.; Lu, X.Y.; Yang, Y.S. Geological characteristics and shale gas distribution of carboniferous mudstones in the Tarim Basin, China. *Acta Geochim.* **2017**, *36*, 260–275. [CrossRef]

39. Cheng, B.; Chen, Z.H.; Chang, X.C.; Chang, X.C.; Yang, F.L. Biodegradation and possible source of Silurian and Carboniferous reservoir bitumens from the Halahatang sub-depression, Tarim Basin, NW China. *Mar. Pet. Geol.* **2016**, *78*, 236–246. [CrossRef]
40. Meng, M.M.; Fan, T.L. Sedimentary characteristics and formation mechanisms of the Ordovician carbonate shoals in the Tarim Basin, NW China. *Carbonates Evaporites* **2020**, *35*, 47–77. [CrossRef]
41. Jiang, D.H.; Yan, S.J.; Pu, R.H.; Su, S.Y.; Zhou, F.; Chen, Q.L. Coal and shale distributions on half-graben slopes: A case study of the Eocene Pinghu Formation in Xihu Sag, East China Sea. *Interpretation* **2022**, *10*, 1–11. [CrossRef]
42. Ashraf, U.; Zhu, P.M.; Yasin, Q.; Anees, A.; Imraz, M.; Mangi, H.N.; Shakeel, S. Classification of reservoir facies using well log and 3D seismic attributes for prospect evaluation and field development: A case study of Sawan gas field, Pakistan. *J. Petro. Sci. Eng.* **2019**, *175*, 338–351. [CrossRef]
43. Mukhopadhyay, P.K.; Wade, J.A.; Kruge, M.A. Organic facies and maturation of Jurassic/Cretaceous rocks, and possible oil-source rock correlation based on pyrolysis of asphaltenes, Scotian Basin, Canada. *Org. Geochem.* **1995**, *22*, 85–104. [CrossRef]
44. Pu, R.H. Geological interpretation of progradational reflection. *Oil Geophys. Prospect.* **1994**, *29*, 490–497.
45. Gu, J.Y. A Study of Sedimentary Environment and Reservoir Quality of the Carboniferous Donghe Sandstone in the Tarim Basin. *Acta Geo. Sin-Engl.* **2009**, *9*, 395–406. [CrossRef]
46. Lüthje, C.J.; Nichols, G.; Jerrett, R. Depositional facies and reconstruction of a transgressive coastal plain with coal formation, Paleocene, Spitsbergen, Arctic Norway. *NJG* **2020**, *100*, 1–49. [CrossRef]
47. Laurent, G.C.; Edoardo, P.; Ian, S.R.; Peacock, D.C.P.; Roger, S.; Ragnar, P.; Hercinda, F.; Vladimir, M. Origin and diagenetic evolution of gypsum and microbialitic carbonates in the Late Sag of the Namibe Basin (SW Angola). *Sediment. Geol.* **2016**, *342*, 13–115. [CrossRef]

Article

Comparison of Axial Flow and Swirling Flow on Particle Pickup in Horizontal Pneumatic Conveying

Yun Ji [1,2,*], Yating Hao [1], Ning Yi [1], Tianyuan Guan [1], Dianrong Gao [1,2] and Yingna Liang [1]

[1] School of Mechanical Engineering, Yanshan University, Qinhuangdao 066004, China
[2] Hebei Provincial Key Laboratory of Heavy Machinery Fluid Power Transmission and Control, Yanshan University, Qinhuangdao 066004, China
* Correspondence: jiyun@ysu.edu.cn

Abstract: Pneumatic conveying is widely used in coal mining. As the lowest conveying velocity of materials, the pickup velocity is the key to the study of gas–solid two-phase flow. In this study, the pickup velocity of pebble particles was experimentally investigated. When the particle size is 3–9 mm, the airflow velocity was found to suitably describe the results as a function of the pickup velocity and have a high correlation. When the swirl number is 0.2, the optimal swirl number was found for which the highest particle pickup ratio was observed. Based on four different methods, namely, visual observation, mass weighing, coefficient of difference analysis, and determination of the peak-average ratio of the pressure drop in the flow field to measure the pickup velocity of the spraying material, the results showed that the accuracy of the particle pickup velocity obtained through visual observation was the lowest, and when the mass–loss rate of the particle was selected as the measurement index of the pickup velocity, the accuracy was the highest. The results will help to realize the long-distance transportation of spraying materials in inclined roadway under the shaft.

Keywords: axial flow; swirling flow; pickup behavior; coefficient of variation

Citation: Ji, Y.; Hao, Y.; Yi, N.; Guan, T.; Gao, D.; Liang, Y. Comparison of Axial Flow and Swirling Flow on Particle Pickup in Horizontal Pneumatic Conveying. *Energies* **2022**, *15*, 6126. https://doi.org/10.3390/en15176126

Academic Editors: Gan Feng, Qingxiang Meng, Gan Li and Fei Wu

Received: 7 July 2022
Accepted: 20 August 2022
Published: 23 August 2022

Publisher's Note: MDPI stays neutral with regard to jurisdictional claims in published maps and institutional affiliations.

1. Introduction

In recent years, the construction technology for coal mining in China has developed rapidly, and the monthly footage of roadway wellbore construction has exceeded 100 m [1,2]. Roadway excavation is of great significance for coal mining. However, the conveying distance of spraying materials is less than 200 m. For a long-distance inclined shaft excavation (≥500 m), conveying the spraying materials by laying the track results in considerable safety risks and low work efficiency. Therefore, long-distance pneumatic conveying of spraying materials is of high economic value and social benefits.

Pneumatic conveying is widely used in chemical, mining, and other industries, which utilizes the airflow to convey particulate material along the direction in a closed pipeline [3,4]. In the past few years, we have devoted ourselves to the study of long-distance pipeline dilute-phase pneumatic conveying of spraying materials, and committed to applying this technology to the long-distance pipeline conveying of underground roadway gangue and coal mining [5,6]. For long-distance pipeline conveying, the pressure drop in the pipeline increases with the conveying distance, resulting in blockage of the pipeline. When the airflow velocity reaches the pickup velocity of particles, particles are disturbed and gradually picked up. Conversely, particles suspended in the pipeline gradually settle and pile up at the pipeline bottom. Researchers have investigated the threshold velocities of pneumatic systems, which can be classified into two groups: the velocity required to move these particles when they are initially moving (pickup velocity) and that to keep the particles moving and prevent them from deposition (saltation velocity).

Pickup velocity is the limit airflow velocity required to pick up the particles [7]. A few studies on pickup velocity have been conducted to reveal the basic features. Soepyan [8]

used a novel combination of various methods to predict the confidence of solid transport models by considering the input conditions, experimental data, and modified model. Rice [9] proposed a general functional form for the critical deposition velocity to analyze the low solid void fractions based on the Reynolds number and Archimedes number of particles. Gomes [10] developed an analytical model to analyze the effect of the particle size and particle shape on particle picking mechanism in a horizontal pipe and found that the pickup velocity increased with a decrease in particle sphericity. Dasani [11] measured the pickup velocity of three glass spheres in water and compared the pickup velocity trends under various particle properties for similar gas-solid phase systems. Kalman [12] presented 100 measurements of the pickup velocity of 24 materials having various particle sizes, shapes, and densities and established simple relationships of three zones between the Reynolds and Archimedes numbers (see Figure 1). However, these studies mainly involved micron-sized powder particles. For pebble particles having a size of more than 3 mm, the saltation and pickup velocity significantly differ from those of the powder. Furthermore, the main focus of the aforementioned studies was the particle pickup regimes, and how to reduce the minimum airflow velocity remains unclear. Reducing the airflow velocity can effectively reduce energy consumption, and how to reduce the particle pickup velocity is still an urgent problem to be solved.

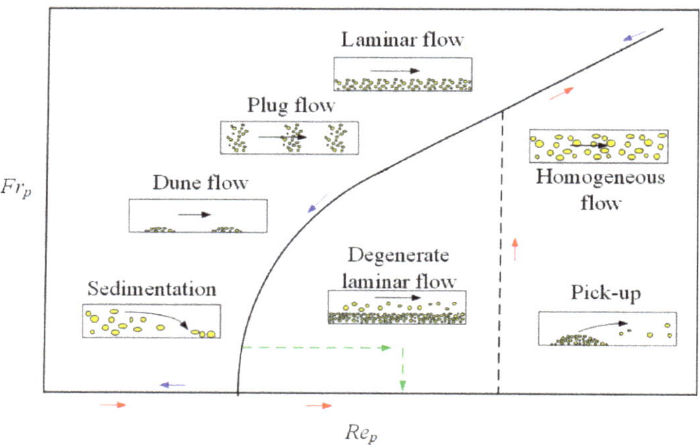

Figure 1. Model of flow regimes observed for horizontal pneumatic conveying.

Swirling pneumatic conveying can effectively reduce the pickup speed of solid particles, and researchers have carried out a lot of research on swirling pneumatic conveying. Zhou [13–15] studied the influence of the type of swirl generator and the intensity of swirl on the pickup speed of lump coal particles, and found that the swirl helps to pick up the particles from the front to the end of the particle bed. Ibrahim [16] used turbulence models to study the swirling pneumatic conveying, and found the most suitable turbulence model in a vertical dryer. Yao [17] studied laminar flow in a horizontal pipeline, focusing on the slug flow and velocity distribution in the pipeline, and studied the effect of the inlet swirl number, Reynolds number, and axial spacing on the swirl decay and velocity distribution. Fokeeret [18,19] studied the three-leaf swirling airflow, obtained the swirl intensity in the pipeline by a laser Doppler anemometer, and studied the effect of the Reynolds number and the distance from the pipeline outlet on the swirl decay. Li [20–24] studied the swirl airflow on vertical and horizontal pipes by comparing experimental results with numerical results. Zhou [25] studied the effects of swirls on gas-solid pneumatic conveying, and determined the tangential velocities of weak and strong swirls. Studies have shown that the swirling pneumatic flow is beneficial for the reduction of the conveying velocity of the solids. However, due to the interaction of the airflow-wall, the swirling flow will lead to the swirl decay, and the coupling effect of the tangential flow and the axial flow on the particle pickup velocity will be unpredictable. Therefore, it is necessary to study the swirling pneumatic conveying for better understanding the particle picking process.

In this study, experiments were conducted to evaluate the pebble particle pickup regimes in axial and tangential flow. The effects of swirl numbers on the axial and swirl velocities in the pipeline were studied, and Gaussian fitting was performed. The airflow velocity was found to appropriately describe all the results as a function of the pickup velocity, and it had a good correlation with the results. For small particles with a diameter of 3–9 mm, Kalman's empirical equation was suitable for the particle pickup velocity, and the pickup velocity increased with the particle size. The highest pickup ratio was observed when the swirling flow number was between 0.1 and 0.2. Based on a comparison of the particle pickup velocities obtained by using four methods, namely, visual observation, mass weighing, analysis of the coefficient of difference (CV) of the pressure drop in a flow field, and determination of the peak-average ratio (PAR) of the pressure drop in a flow field, the accuracy of the particle pickup velocity obtained through visual observation was found to be the lowest, and the highest was obtained by considering the accuracy of the particle mass-loss rate as the measurement index of the pickup velocity. For pipeline pneumatic conveying, where measuring the material pickup rate is difficult, the particle pickup velocity and optimal swirling flow number can be obtained by determining the airflow pressure characteristics of the flow field, CV of the pressure signal, and PAR.

2. Experimental Setup

2.1. Experimental Materials

For particles having fixed sizes, the higher the particle density, the higher the pickup velocity. To simplify the experiment and facilitate the observation of particle flow patterns, six particles were selected by seven round screens with apertures of 3 mm, 5 mm, 7 mm, 9 mm, 11 mm, 13 mm, and 15 mm, and their sizes were 3–5 mm, 5–7 mm, 7–9 mm, 9–11 mm, 11–13 mm, and 13–15 mm, respectively. Figure 2 and Table 1 showed the results of the slump experiment of pebble particles, which were released on the surface of an acrylic plate, and the angle of repose of the pebble particle group varied from 39° to 35°. Due to the random distribution of particle size, the structure of particles was asymmetric. Larger particle sizes would result in greater angles of repose. However, the particles gradually tended to be smooth with the decrease of the size. This was because small particles accumulated more densely, and they presented better symmetry.

2.2. Experimental Apparatus

Figure 3 shows a schematic of the experimental apparatus, which were used to study the picking process of pebble particles. In the experiments, a screw air compressor was used as the source of airflow supply with 6.9 m^3/min, and its loading and unloading pressures were 0.4 MPa and 0.42 MPa, respectively. The maximum experimental airflow velocity was 200 m^3/h. The conveying pipelines used in this experiment was 50 mm; to avoid blockage in pipelines, the internal size of the pipelines should be at least three times the size of the pebble particles. An air tank was also arranged in the pipeline. In addition, it could reduce the unstable airflow generated during the loading and unloading processes. To measure the steady airflow, the stable length of the flowmeter should be at least 50 times the size of the pipeline. Therefore, the distance between the spherical valve and flowmeter was set to 70 D_{pl} in this experiment. The swirling airstream was generated when the tangential airflow passed through the swirling generator, and the particles deposited at the bottom of the pipe were disturbed at this time. The pressure sensor was installed in front of the test pipeline, and its pressure was considered as the pressure drop of the pipeline because the outlet of the pipeline was a free flow outlet, and the gauge pressure at the outlet part was approximately zero.

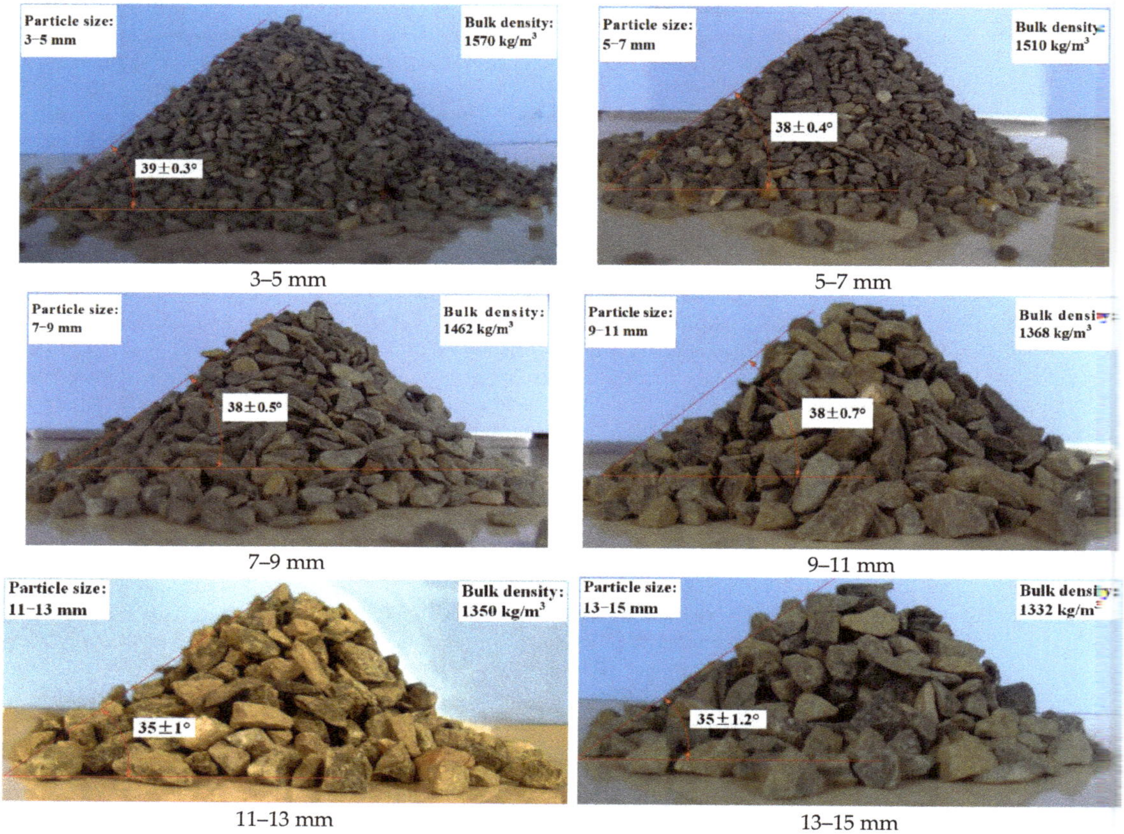

3–5 mm

5–7 mm

7–9 mm

9–11 mm

11–13 mm

13–15 mm

Figure 2. Slump test of pebble particles in the experiments.

Table 1. Particle sizes and their properties.

Particle size (mm)	3–5	5–7	7–9	9–11	11–13	13–15
Angle of repose (°)	$39 \pm 0.3°$	$38 \pm 0.4°$	$38 \pm 0.5°$	$38 \pm 0.7°$	$35 \pm 1°$	$35 \pm 1.2°$
Bulk density (kg/m³)	1570	1510	1462	1368	1350	1332

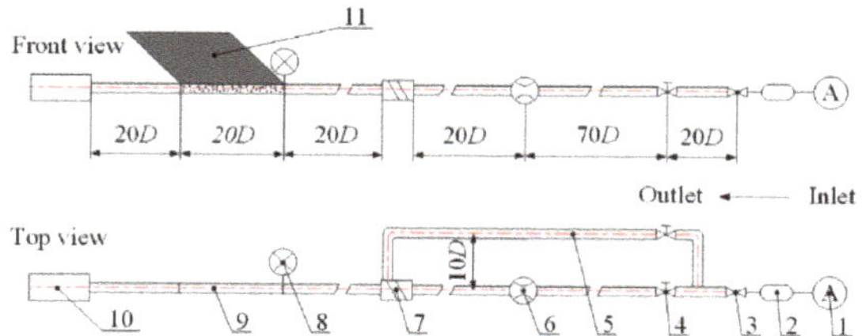

Figure 3. Schematic figure of the experimental setup. 1—Air compressor, 2—gas tank, 3—spherical valve, 4—spherical value, 5—tangential flow, 6—flowmeter, 7—swirling generator, 8—pressure sensor, 9—acrylic transparent tube, 10—receiving device, 11—high speed camera.

Figure 4 showed a cross-sectional view of a swirling generator. Three swirling blades were arranged to avoid the swirl decay in the case of too many blades. By changing the regulating sleeve, the swirling number of the swirling flow could be adjusted accordingly. In the experiments, the total airflow varied from 40 m^3/h to 180 m^3/h. The typical ratio between the gas nozzle and the main inlet of the swirling generator was the ratio between the gas flow into the nozzle 12 as shown in Figure 4 and the total airflow. In each group of experiments, we kept the total inlet flow fixed. By changing the typical ratio, we could achieve different swirling numbers in the pneumatic conveying. The swirling intensity was adjusted by adjusting the size of the tangential flow valve, and the tangential flow ratio varied from 0–1 (see Table 2).

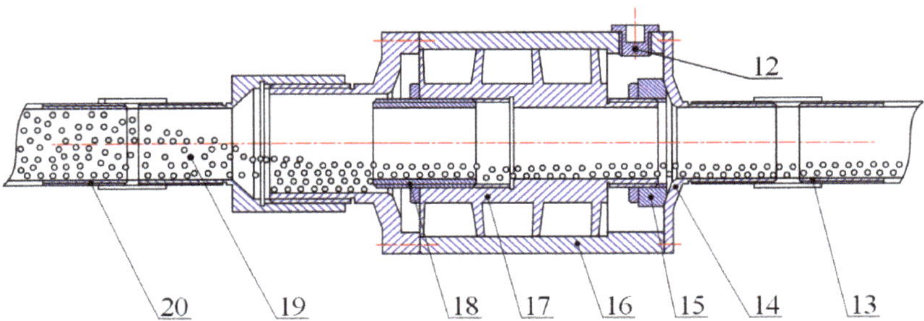

Figure 4. Split view of swirling generator [6]. 12—Gas nozzle, 13—inlet pipe, 14—right cover, 15—adjusting ring, 16—housing, 17—swirling blade, 18—regulating sleeve, 19—pebble particles, 20—outlet pipe.

Table 2. Experimental setup.

Value	Particle Size (mm)	Airflow (m^3/h)	Tangential Flow Ratio
Numerical value	3–5, 5–7, 7–9, 9–11, 11–13, 13–15	40–180	0–1

A vortex flowmeter and weighing sensor were used to achieve the airflow rate and particle mass, respectively, and the swirl number (S_N) could be defined as [14,15,21,23]:

$$S_N = \frac{\int_0^R 2\pi \rho_a u w r^2 \mathrm{d}r}{\int_0^R 2\pi \rho_a R u^2 r \mathrm{d}r} \tag{1}$$

where u and w are the axial and tangential velocities, respectively, and r is the radius of the pipe.

The experiments were implemented in a horizontal transparent pipe having a length of 1 m, the total mass of pebble material used in each experiment was 500 g, and the pebble material was tiled horizontally at the bottom of the pipeline having a tiling length of 500 mm. A fixed airflow rate was applied to the granular layer through the inlet of the pipeline. The particles accumulated in the transparent acrylic tube changed from static to rolling, sliding, or suspended under the airflow. The initial motion state and pickup process of pebble particles were recorded by a high-speed camera namely JVC/gc-px100bac, with a 500 FPS high-speed film recording capacity, and 500 FPS was used for experiment recording. In the experiments, the particles picked up by the airflow were collected through a porous filter that allows the airflow to pass but the particles. The masses of the picked particles were obtained by an electronic balance with a range of 2 kg, and its resolution was 0.01 g. Finally, the pickup velocity of the particles was calculated according to the 50% mass loss of pebble particles under different airflow rates.

3. Results and Discussion

3.1. Airflow Distribution

The airflow velocity is one of the key parameters that affect the flow pattern of the two-phase flow, and it is important to study the airflow distribution in horizontal pipelines.

Figure 5 shows the measurement area of the axial and tangential velocity distribution in the horizontal pipe; and the area was divided into five spaces from the top to the bottom. The axial airflow velocity was that along the positive direction of the z-axis, and the tangential airflow velocity is the positive direction along the y-axis. To reduce the errors caused by accidental experiments, the airflow velocities were measured thrice, and their average value was taken as the final result.

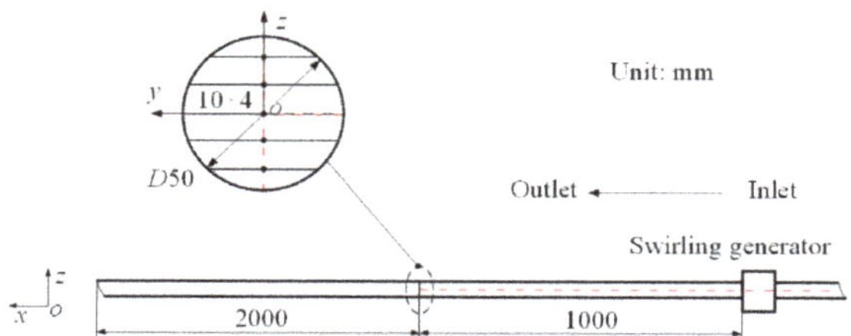

Figure 5. Schematic of experimental measuring domain.

Thermal anemometer AR866 was selected as the anemometer in the experiments, and its measuring range was 0.3–60 m/s with a resolution of 0.01 m/s. The measured airflow rates were those under standard conditions. To measure the axial airflow velocity, a flow of 200 m^3/h under standard conditions was chosen as the velocity measurement target.

Figure 6 shows the airflow distribution of the axial velocity (z-direction), and the color lines are the Gaussian fitting for different swirl numbers. For different swirl intensities, the axial flow velocities in the lower part of the pipeline are larger, whereas those in the upper part are smaller, which are attributed to gravity and the swirling generator. The axial speed of the flow increases first and then decreases with the swirl numbers. When the swirl number is $S_N = 0.3$, the axial speed reaches a maximum value of 41 m/s. With the increasing swirl number, the axial velocity of the pipe gradually decreases. This is because the axial velocity decreases with the swirl number, whereas the total flow remains unchanged, and the friction between the swirl flow and the wall intensifies. Compared with the axial flow, the swirl decay is more significant, and thus, the axial velocity decreases.

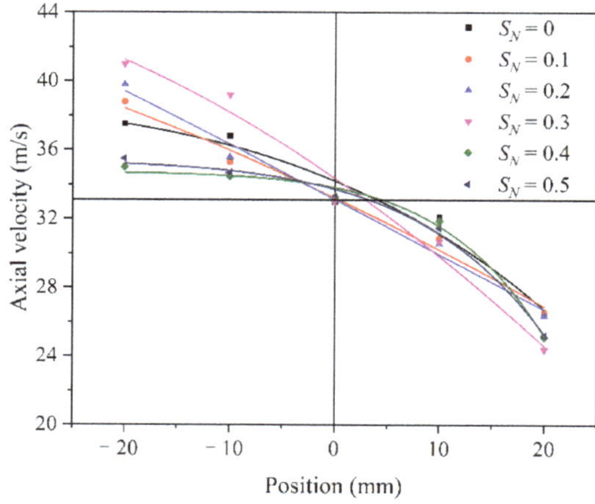

Figure 6. Axial airflow distribution and Gauss fitting under different swirl numbers.

Figure 7 shows the distribution of the tangential airflow velocity (y-direction), and the color lines are the Gaussian fitting for different swirl numbers. The airflow velocity along the positive y-axis is the positive direction, and the negative direction indicates that the airflow velocity is along the negative the *y*-axis. The trends of the tangential and axial velocities are the same, and the tangential velocity in the upper part of the pipe is significantly smaller than that in the lower part of the pipe. The swirling generator plays an important role in airflow distribution.

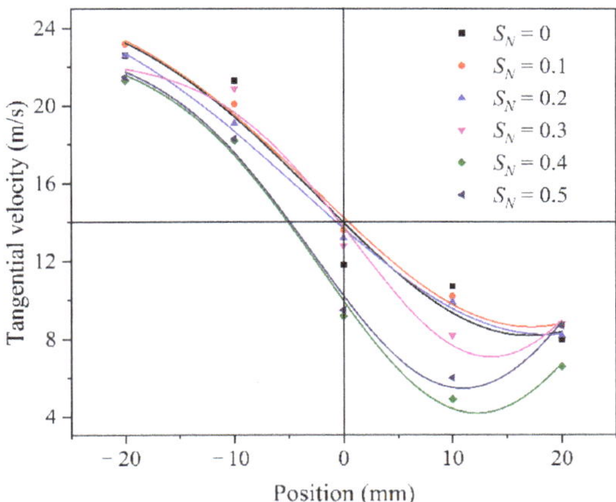

Figure 7. Tangential airflow distribution under different swirl numbers.

Furthermore, Figure 7 shows that with the increasing swirl number, the tangential velocity of the airflow increases gradually. However, the tangential velocity reaches a maximum with a swirl number of 0.3, indicating that when the swirl number is 0.3, the tangential swirl intensity in the horizontal pipe is the highest. It is found that the swirl number plays an important role in the swirling decay, the larger the airflow velocity is, the greater the swirl decay is, and the tangential airflow velocity will decrease accordingly. When the swirl number reaches 0.5, a lowest tangential airflow velocity is found. Simultaneously, the growth trend of the velocity near the wall of the pipeline is weakened, and presents a downward trend. This is because the velocity near the wall is less than that at the central axis owing to the friction between the wall and airflow.

3.2. Particle Pickup in Axial Flow

The experiments were implemented in a horizontal transparent pipe having a length of 1 m; the total mass of the pebble material in each test was 500 g, and was tiled horizontally at the bottom of the pipeline having a tiling length of 500 mm. A fixed airflow rate was applied to the granular layer through the inlet of the pipeline. The granular layer was picked up by airflow entrainment, and the surface airflow velocity was larger than the inlet velocity. As the particle layer eroded and gradually picked up, the average airflow velocity at the particle surface layer decreased (the cross-sectional area of the airflow increased), and the airflow velocity at the particle surface was calculated by dividing the volume airflow rate of the pipeline (free cross-sectional area), as shown in the following equation:

$$v_{sf} = v_{in} \frac{\pi r^2}{\pi r^2 - \frac{m_e}{\rho_b L_e}} = v_{in} \frac{\pi \rho_b L_e r^2}{\pi \rho_b L_e r^2 - m_e} \tag{2}$$

where v_{sf} is the airflow velocity on the particle surface, v_{in} is the inlet airflow velocity, m_e is the total mass of particles in each test, L_e is the particle layer length that were tiled, and ρ_b is the particle bulk density.

In the picking process of the 3–5 mm particles, the airflow at the inlet was 137–140 m³/h, and the picking process of pebble particles in the horizontal pipe was recorded by using a high-speed camera, as shown in Figure 8. When t = 0.00 s, the particles remain still.

With time, the particles on the surface of the particle layer are gradually picked up. When $t = 0.05$ s, two protruding particles are picked up. The main reason is that the windward side of the particles is considerably large, and the horizontal force on the particles increases. When the force of airflow exceeds the friction force of the particles, the particles are picked up, begin to roll, and picked up again. As the laying of particles at the bottom of the pipeline is significantly affected by random factors, and the friction between the particle and wall, friction between particle and particle, particle size, and particle bulk density has a certain effect on the picking of pebble particles, the particles sometimes stop moving again after being picked up, as shown in Figure 8b. In this stage, only the particles with an uneven surface are forced more in the wind, and few particles are picked up. Therefore, to avoid the random influence of these factors on the picking of pebble particles, each test was performed twice. If the mass error of the two tests exceeded 20%, the test was conducted three or even four times. The singular test results were filtered out, and the average value of several groups of test picking was obtained.

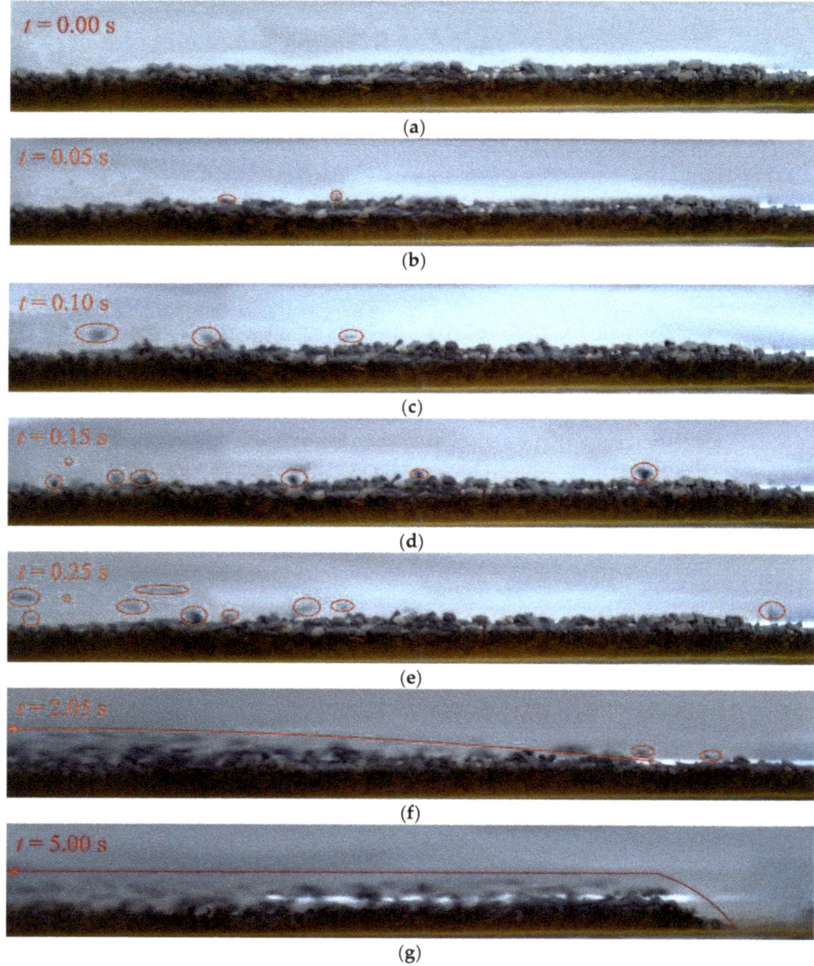

Figure 8. Pick-up process of pebble particles. (**a**) $t = 0.00$ s, (**b**) $t = 0.05$ s, (**c**) $t = 0.10$ s, (**d**) $t = 0.15$ s, (**e**) $t = 0.25$ s, (**f**) $t = 2.05$ s, (**g**) $t = 5.00$ s.

Figure 8 depicts that particle-picking starts with a few particles rolling and impacting other particles, and the particles at the end of the pipeline are picked up gradually and suspended. When the particles are reduced to a certain number, the particles move in the form of laminar flow. The picking process can be divided into four stages. First, particles begin to roll along the particle layer (Figure 8b). Second, the picked particles collide with other particles in the layer (Figure 8c–e), and the particles begin to rebound. Third, the newly picked particles collide with the particles already carrying air (Figure 8f). Fourth, the windward side of the particle group is picked up gradually, and the particle-picking process ends.

It can be seen from Figure 8c–f that as few particles are picked up, the picked-up particles roll and jump, thereby impacting the deposited particles, and these particles are picked up (as shown in Figure 8c–e). The picked-up particles impact other particles in the rolling process, causing the particles on the surface of the particle group to be picked up gradually. The picking process of the particle group gradually slows down with cutting, as shown in Figure 8f. With the picking process, the number of particles deposited at the bottom gradually decreases, and the friction between the particles and the wall decreases. When the fluid force on the windward side of the particles exceeds the friction of the particle pile, the particles move along the outlet in the form of laminar flow, as shown in Figure 8g.

Some randomness exists in the particle-picking process; thus, the particle-picking regimes will appear in a limited range of airflow velocity only. Therefore, the pickup velocity of the particles determined by visual observation will lead to a certain error, and a reasonable pickup velocity is a key to the research on the process of particle-picking in horizontal pipes; some researchers had a different definition of pickup velocity in the past. To solve the problem of micro-particle dune flow, Hayden et al. [26] slanted the material inside the pipe and made it away from the air source. They inferred the pickup velocity of particles inversely by measuring the height of the stabilized particle group and made the airflow velocity when the mass of particles picked up was larger than zero be the pickup velocity of materials. Kalman et al. [12] defined the mass loss of particles within 30 s in one test, and the pickup velocity was identified as the airflow velocity when the material mass loss was not zero. Rabinovich [27] and Zhou [12] defined the airflow velocity when the picking mass of the particle was 50% as the pickup velocity. To reduce the influence of random factors on the pickup velocity, this paper tended to measure the pickup velocity qualitatively, the pickup velocity of pebble particles was determined by drawing the function relationship between the airflow and the entrainment of particles, measuring the intersection point of measurement point and vertical coordinate as well, the pickup velocity was treated as the airflow velocity when the picking mass was 50%. Based on the function relation of the mass loss, higher measurement accuracy could be obtained. The greater the number of particles measured, the higher the measurement accuracy.

Figure 9 shows the functional relationship between the pickup velocity of pebble particles with different sizes and airflow velocities. The picking rate is selected as the measurement index of the pickup velocity, that is, the ratio of the picking mass to the total mass of the material. The pickup ratios are low at low airflow velocity for different sizes due to the solid pickup velocity, and then a large number of particles are picked up in a small velocity range. For pebble particles with different particle sizes, the 50% pickup velocities are 17.9 m/s, 20.5 m/s, 22.1 m/s, 20.7 m/s, 17.0 m/s, 19.5 m/s, respectively. When particle pickup begins, the picking rate of the particles has an approximately linear relationship with the airflow velocity. For various sizes of pebble particles, the bulk density decreases with the particle size, resulting in the increasing of the porosity. The larger the particle size, the higher the accumulated height of the particles, which will increase the surface airflow velocity of the particles. Rabinovich et al. [27] found that the maximum pressure drop above the particle accumulation is equal to 10% when the accumulation at the bottom was less than 70%. They considered that the effect of particle accumulation on particle-picking can be ignored. However, when the sediment height exceeds 90% of the pipeline, the calculation of the pickup velocity may lead to errors that cannot be ignored. The effects of the airflow velocity, material permeability, and particle stacking height should be considered.

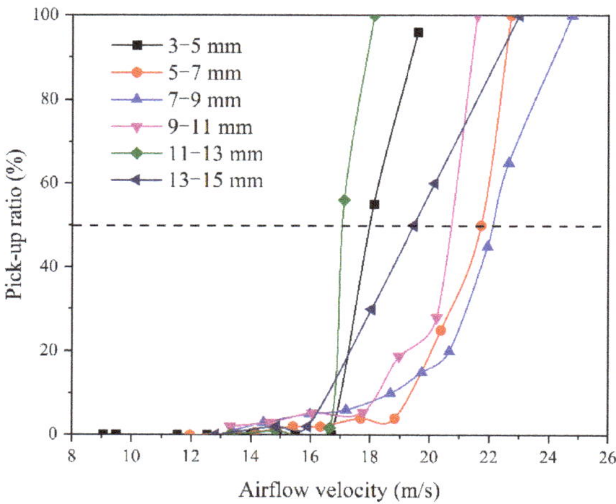

Figure 9. Pick-up ratio of pebble particles with different particle sizes.

Figure 10 shows the pickup velocity of pebble particles with a picking rate of 50% for particles with different particle sizes. For particles from 3 to 9 mm, the pickup velocity increased with the increasing size, which is consistent with the research results of Kalman, Rabinovich, and Zhou et al. [12,27]. As regards the size of particles in a specific range, Kalman's empirical equation was more suitable for the prediction of the particle pickup velocity of the particles considered in this study. For particles ranging from 9–11 mm, the surface airflow velocity and the height of the windward side had significant effects on particle-picking, and the pickup velocity of the particles was low. For pebble particles of 5–9 mm, the pick-up masses were the minimum. Therefore, for pneumatic conveying of the gas-solid two-phase flow, it was often required that the particle size was less than three times the pipe diameter. When the particle size was close to 1/3 of the pipe diameter, the particle would not be suitable for conveying in such a small pipe.

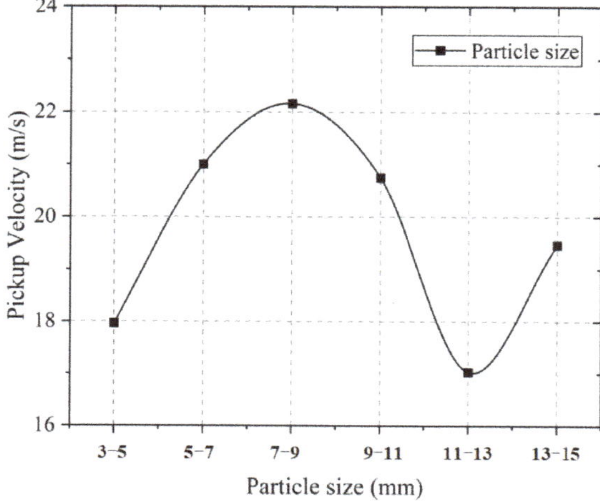

Figure 10. Function between particle size and pick-up velocity.

The stress analysis of a particle in the axial flow field of a horizontal pipeline is shown in Figure 11. In the figure, u_a is the airflow velocity vector, u_p and ω_p are the velocity the vector and angular velocity vector of spraying particles respectively. According to the gas–solid two-phase hydrodynamics, the forces received by spraying particles when moving in the flow field include the gravity, the drag force generated by the relative velocity between airflow, the lift force generated by the particles rotation, and pressure gradient force generated by the velocity difference between the upper side and lower side of the particles and pressure gradient force, buoyancy force and other forces generated by temperature difference and electric field. The above forces have different effects on particle motion in different airflow. In the pneumatic conveying system of spraying particles, there is a great difference between the particle density and the gas density, and the size of a single spraying particle is large. Among the above forces, only the drag force, rotating lift force and particle inertia force are of the same order of magnitude and have an impact on the movement.

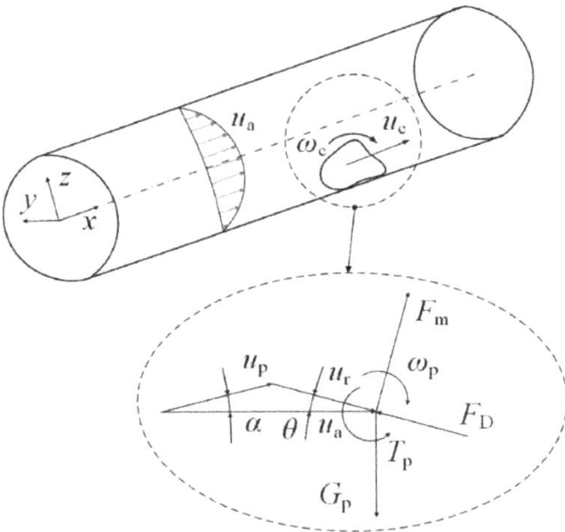

Figure 11. Forces analysis of single spraying particle in horizontal axial pneumatic conveying.

When measuring the dispersion of different data, directly comparing the two groups of data with the standard deviation will lead to a large error. Comparing with CV can eliminate the influence of measurement scale and size. The CV is the ratio of the standard deviation of the original data to the average value, as shown in Equation (3). There are no dimensions, so objective comparisons can be made.

$$\mathrm{CV} = \frac{\sigma_x}{E_x} \tag{3}$$

where CV is the coefficient of variation, σ_x is the standard deviation, E_x is the average.

Figure 12 shows the normalized functional relations between the airflow velocity and the CV of the pressure under different particle sizes. In this study, because the end of the test pipeline was a free flow outlet, the pressure of the pressure transmitter could be regarded as the pressure drop of the tested pipeline. CV can be used to describe the fluctuation of the flow field in a horizontal pipeline. The larger the CV, the more unstable the flow field. For particle multiphase flow, particle pickup and suspension affect the stability of the flow field. Therefore, CV is selected as another index to measure the pickup velocity.

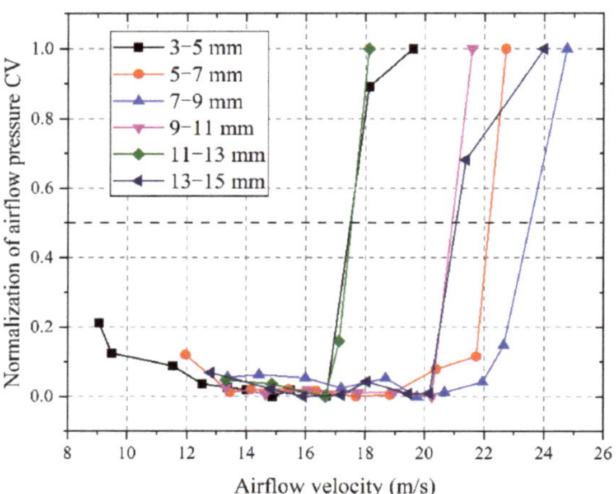

Figure 12. Relationship between coefficient of difference and airflow velocity.

For lower airflow velocity, the CV of the airflow pressure is low, which is less than 0.2. This is mainly because the airflow velocity is less than the particle pickup velocity, so only a few particles are picked up at this low airflow velocity. The fluctuation of the flow field is caused by the decrease in the ventilation cross-sectional area of the flow field caused by the particle pile, and the periodic acceleration and deceleration of the airflow caused the fluctuation of the gas. With the increasing airflow velocity, the particles begin to be picked up, and the pressure CV increases gradually. For the pebble particles of 3–9 mm, with the increasing particle size, the airflow velocity will be larger, which is similar to the result of the picking rate measurement index. Furthermore, no linear relationship exists between the CV of the pressure and particle size with sizes of 9–11 mm, 11–13 mm, and 13–15 mm. When the particle size exceeds 9 mm, the particle pickup velocity is smaller than that of the 7–9 mm particles. One reason is that with the increasing particle size, the porosity of particles increases, the packing density of particles decreases, the permeability of particles increases, and large particles tend to be disturbed and picked up by airflow. In addition, for horizontal pneumatic conveying with fixed pipe diameter, the surface airflow velocity of the large particles is higher; thus, the pickup velocity will be smaller.

Some errors may exist when the particle pickup velocity is measured using the CV of a flow field if the time of particle-picking is substantially long or the particle-picking process may be completed in an instant or within a period. Therefore, the PAR is adopted to further optimize the pickup velocity obtained by the CV in this study, as shown in the following equation:

$$r_{par} = \frac{x_{i,\max}}{E_X} \tag{4}$$

where r_{par} is the peak-average ratio of the pressure drop, $x_{i,\max}$ is the peak pressure signal, and E_x is the average of the pressure drop.

Figure 13 shows the normalized function between the airflow velocity and PAR under the pipeline pressure of particles of different sizes, it can be seen from the figure that the relation of PAR has a certain similarity with the function of CV, the pickup velocities obtained by taking 50% of the PAR as the calculation are 17.54 m/s, 22.17 m/s, 22.51 m/s, 20.95 m/s, 17.57 m/s, and 22.74 m/s respectively, which are approximately equal to the pickup velocity obtained by taking the CV as the measurement index in Figure 13. For particles of 13–15 mm, the pickup velocity obtained by taking the PAR as the measuring index is higher than those obtained by using the CV, and the pickup velocity is closer to that obtained by using the picking volume, therefore, the pickup velocity of the particles can be accurately obtained by using the PAR within a certain range.

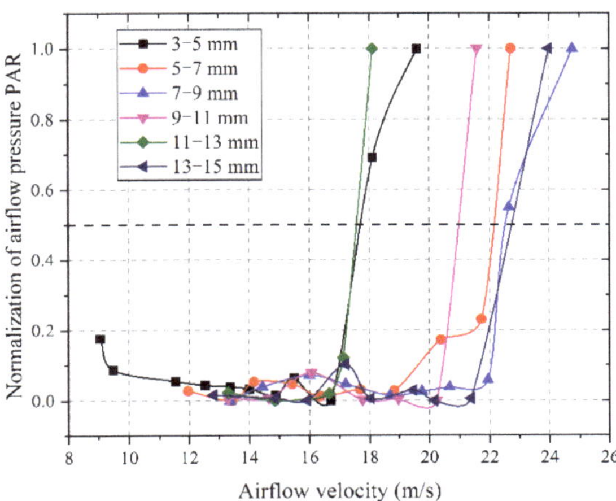

Figure 13. Relationship between peak-to-average rate and airflow velocity.

By comparing Figures 9, 12 and 13, the airflow velocity present the same effect for particle pickup mass, airflow pressure CV, and PAR. In other words, for pipeline pneumatic conveying which is not conservative for the measurement of the material pickup rate, the particle pickup velocity, and optimal swirl number can be obtained by determining the pressure characteristics of the flow field, CV of the pressure signal, and PAR. When the pickup quality is used to evaluate the pickup velocity, the accuracy is the highest, the pressure CV is the second, and par is the worst.

3.3. Particle Pickup in Swirling Flow

It was found from the results of the pure axial flow experiment that the pickup velocity of particles with a particle size of 3–9 mm exhibited a linear relationship. Therefore, pebble particles with particle sizes of 3–5 mm, 5–7 mm, and 7–9 mm were selected as the test objects for flow transportation in the swirling flow field, as shown in Equation (1); when the total flow remains constant as the premise, by changing the swirl number, the pickup velocity of pebble particles under different swirl numbers can be obtained, as shown in Figure 14.

As seen in Figure 14a, for pebble particles of 3–5 mm, when the total airflow velocity is lower than 17.54 m/s, the S_N has an insignificant effect on the particle pickup velocity and instead has an inhibiting effect. With the increasing inlet airflow velocity, a relatively lower swirl number promotes the particle-picking process. When the swirl number is between 0.1 and 0.2, the particle pickup rate reaches the maximum, whereas, with the increasing swirl number, the particle pickup rate decreases gradually owing to the swirl decay. For particles having sizes of 5–7 mm and 7–9 mm and low airflow velocity, as shown in Figure 14b,c, the swirl number has no significant effect on the pickup rate; however, when the airflow velocity is almost similar to the pickup velocity, the swirl number seriously affects the particle-picking ratio. When the swirl number is between 0.1 and 0.2, the pickup rate of particles is the largest, which is similar to the pebble particle swirl conveying with a size of 3–5 mm, that is, the best swirl number exists, and the pickup rate of materials is the largest.

For the swirling pneumatic conveying, the spin motion of spraying particles in the transportation process can be ignored, but the spiral motion of particles around the pipeline axis cannot be ignored. The stress analysis of a single particle in the horizontal pipeline swirling field is shown in Figure 15. In the axial direction, the spraying particle is only affected by the axial drag of the airflow, and in the tangential direction, the particle is affected by gravity, drag, and centrifugal force.

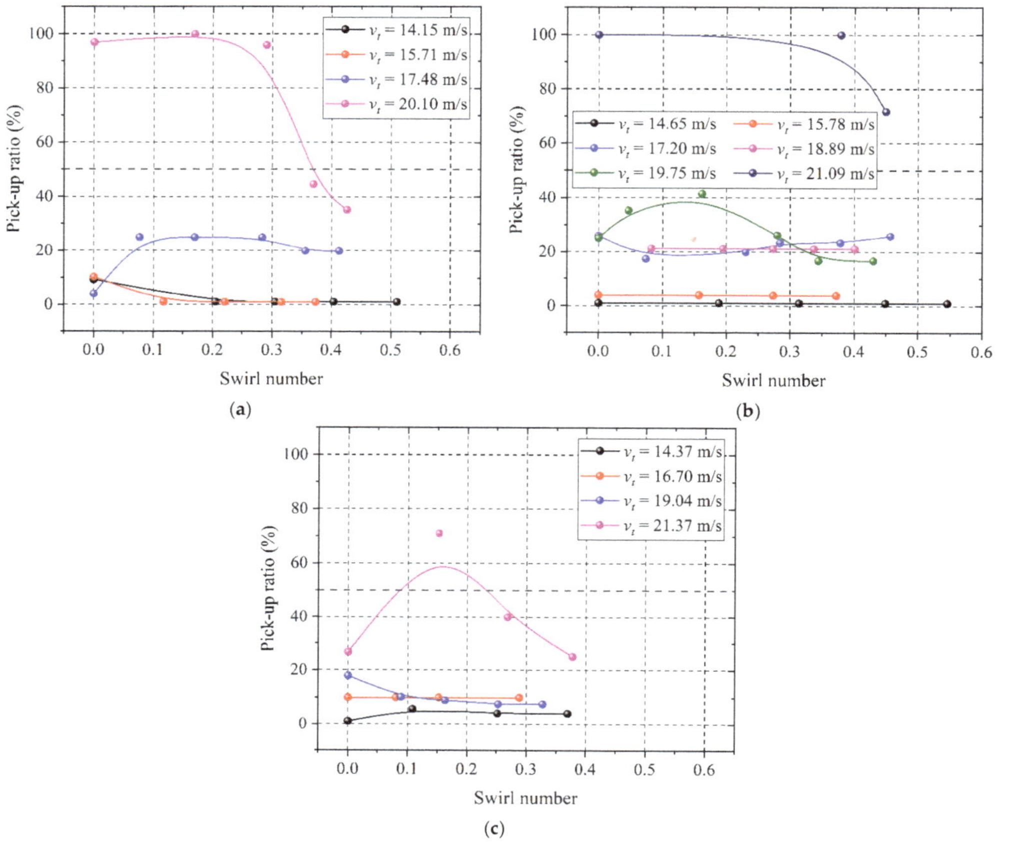

Figure 14. Effect of swirl number on picking velocity of pebble particles. (**a**) d_p = 3–5 mm, (**b**) d_p = 5–7 mm. (**c**) d_p = 7–9 mm.

For pebble particles with large sizes, the pickup velocity differs significantly from that of small particles, mainly because large particles have better air permeability, lower pack-density between particles, and a more significant disturbance effect of swirling flow on the particle group, making it easier for the particles to be picked up. A larger swirl number is not conducive to particle pickup, which is mainly because of the swirl decay. Increasing the swirl number means reducing the axial airflow velocity. Although the tangential flow is conducive to particle disturbance, there will be an optimal swirl number in the process of particle pickup due to the large friction loss energy between the tangential flow and the pipe wall.

Therefore, the swirling velocity can be estimated by using the following equation:

$$v_{pu,s} = \left[-a_s \left(x - S_{N,op} \right)^2 + b_s \right] v_{pu}, \; v_{pu} < f \left(d_{p,i}/d_{p,1} \right) v_{in} \tag{5}$$

where $v_{pu,s}$ is the particle pickup velocity under the swirling flow, a_s and b_s are constants associated with conveying materials, and they are all larger than 0, and $S_{N,op}$ is the swirl number corresponding to the maximum pickup rate. For the pebble particles investigated in this study, $0.1 < S_{N,op} < 0.2$ $f \left(d_{p,i}/d_{p,1} \right)$ is a function related to the particle size, and v_{in} is the inlet airflow velocity.

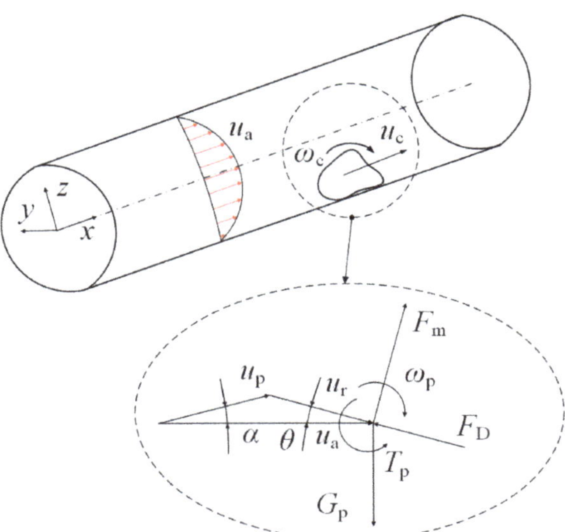

Figure 15. Forces analysis of single spraying particle in horizontal swirling pneumatic conveying.

Figure 16 shows the normalization between the airflow velocity and the pressure CV of the swirling pneumatic conveying. It can be seen from the figure that when the airflow velocity was smaller than the particle pickup velocity, the CV of the flow pressure was relatively low, and they were all less than 0.2. With the increasing swirl number, the CV decreased gradually. This indicated that the particle on the flow field was progressively weakened, and only a small amount of particles were picked up. For pebble particles with a grain size of 3–9 mm, when the airflow velocity was close to the pickup velocity, the particles began to be picked up with the swirl number, and it reached the maximum when the swirl number was 0.1–0.3. With continuously increasing the swirl number, the CV of flow pressure drop decreased due to the swirl decay. Compared with the measurement index of pickup rate, the results of the CV had a certain similarity with the former, which meant that the particle pickup regimes can also be obtained qualitatively by measuring the CV of airflow pressure in horizontal pipelines.

Figure 17 shows the normalization between the airflow velocity and the PAR. The PAR has a high similarity with the pickup velocity obtained from the particle pickup rate results and CV results, and the PAR reaches its maximum value at a swirl number of 0.1–0.3. The PAR of pressure in the flow field is highly reliable as the pickup index of swirling pneumatic conveying, the pickup velocities obtained by the PAR as the measuring index are more accurate than the CV, and the pickup velocity is closer to those obtained by pickup rate.

Comparing the pure axial flow field with the swirl flow field, the accuracy of particle pickup velocities obtained by visual observation is the worst, and it reaches the best when the particle mass loss rate of selecting is selected as the index. For particle conveying which is not convenient to measure the particle pickup ratio, the particle pickup velocity, and the optimal swirl number could be obtained qualitatively by measuring the airflow pressure drop, the CV of pressure signals, and the PAR.

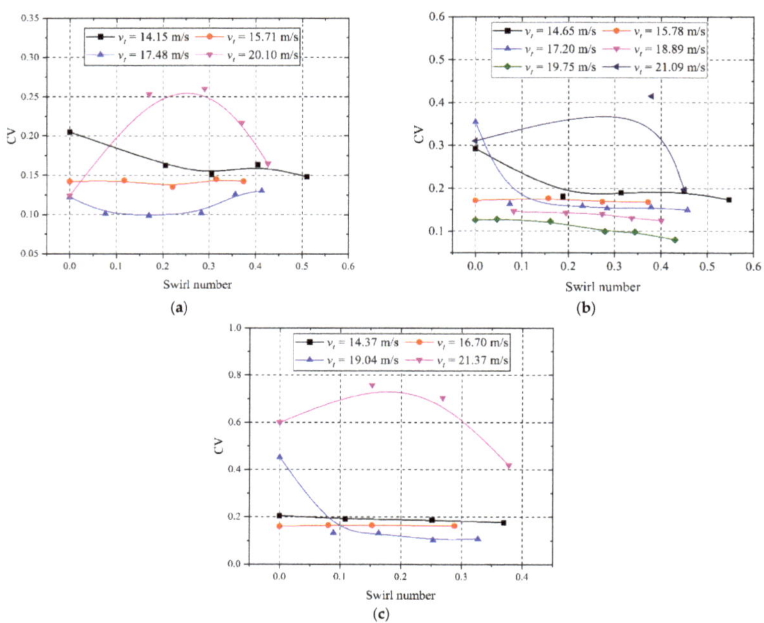

Figure 16. Effect of swirl number on coefficient of variation of pebble particles. (**a**) d_p = 3–5 mm, (**b**) d_p = 5–7 mm, (**c**) d_p = 7–9 mm.

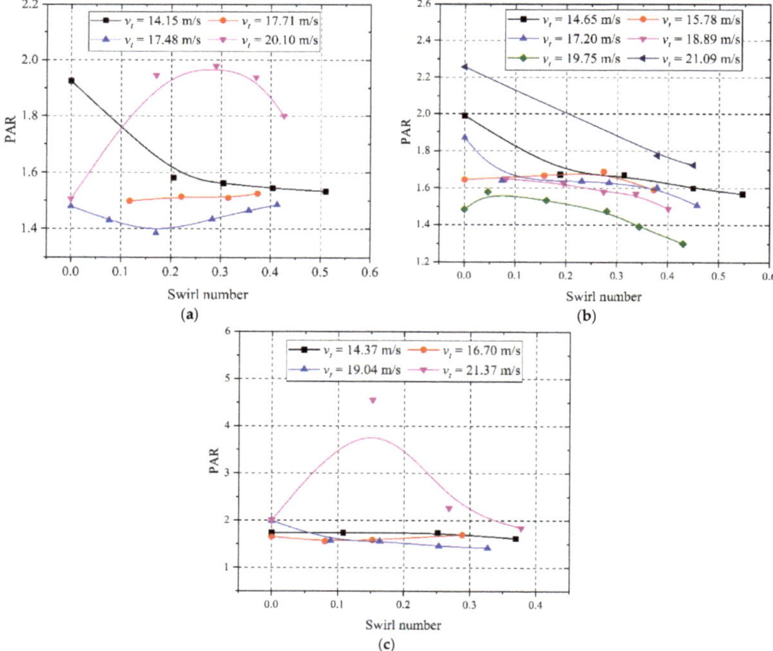

Figure 17. Effect of swirl number on peak to average ratio of pebble particles. (**a**) d_p = 3–5 mm, (**b**) d_p = 5–7 mm, (**c**) d_p = 7–9 mm.

4. Conclusions

The pickup velocity of the pebble particles selected in this experiment was studied based on an empirical analysis of the particle pickup velocity and existing research results. The airflow velocity is found to appropriately describe all the results as a function of the pickup velocity, and a good correlation exists between them. Furthermore, it is found that the experimental object was suitable for the Kalman empirical equation when the range of the particle size is 3–9 mm, and the pickup velocity increased with the particle sizes. Large particles have larger windward surface areas and voidage, and their pickup velocity is affected by the particle size and the internal size of the pipeline.

For pebble particles with large sizes, the pickup velocity is significantly different from that of small particles, and this is mainly attributed to the influence of the windward surface area and particle voidage of large particles. When the swirling flow number ranges from 0.1–0.2, the particle-picking rate reaches the maximum. When the particle pickup velocities obtained by using four methods, namely, visual observation, mass weighing, analysis of the CV of the pressure drop in a flow field, and determination of the PAR of the pressure drop in the flow field, were compared, the results showed that the accuracy of the particle pickup velocity obtained through visual observation was the lowest, and that obtained by selecting the particle mass-loss rate as the measurement index was the highest. For pipeline pneumatic conveying, which is not convenient for the measurement of the material pickup rate, the particle pickup velocity and optimal swirl number can be obtained by determining the pressure characteristics of the flow field, CV of the pressure signal, and PAR.

Author Contributions: Y.J.: conceptualization, project administration, writing—original draft, funding acquisition; Y.H.: writing—review and editing, methodology; N.Y.: investigation, formal analysis; T.G.: investigation; D.G.: writing—review and editing; Y.L.: writing—review and editing. All authors have read and agreed to the published version of the manuscript.

Funding: This research was funded by National Natural Science Foundation of China (grant number 52005430) and Natural Science Foundation of Hebei Province (grant number E2021203108).

Institutional Review Board Statement: Not applicable.

Informed Consent Statement: Not applicable.

Data Availability Statement: Not applicable.

Acknowledgments: This study was supported by the National Natural Science Foundation of China (Grant No. 52005430), and the Natural Science Foundation of Hebei Province (Grant No. E2021203108), which are all gratefully acknowledged.

Conflicts of Interest: No potential conflict of interest was reported by the authors.

Nomenclature

S_N	Swirl number	-	u	Airflow axial velocity	m/s
R	Pipe radius	m	w	Airflow tangential velocity	m/s
ρ_a	Gas density	kg/m^3	v_{sf}	Particle surface airflow velocity	m/s
v_{in}	Inlet airflow velocity	m/s	m_e	Each test particle mass	kg
ρ_b	Particle bulk density	mm	L_e	Particle layer length	m
CV	Coefficient of variation	-	E_x	Mean value	kPa
σ_x	Standard deviation	kPa	R_{par}	Peak-average ration	kPa
$X_{i,max}$	Peak pressure	kPa	$v_{pu,s}$	Swirl pickup velocity	m/s
a_s	constant	-	b_s	constant	-
v_{pu}	Pickup velocity	m/s	$S_{N,op}$	Maximum pickup swirl number	-
r_{xx}	Auto-correlation function	kPa	n	Sequence length	-

References

1. Zhang, Z.; Zhang, R.; Cao, Z.; Gao, M.; Zhang, Y.; Xie, J. Mechanical Behavior and Permeability Evolution of Coal under Different Mining-Induced Stress Conditions and Gas Pressures. *Energies* **2020**, *13*, 2677. [CrossRef]
2. Cheng, L.; Ge, Z.; Chen, J.; Ding, H.; Zou, L.; Li, K. A Sequential Approach for Integrated Coal and Gas Mining of Closely-Spaced Outburst Coal Seams: Results from a Case Study Including Mine Safety Improvements and Greenhouse Gas Reductions. *Energies* **2018**, *11*, 3023. [CrossRef]
3. Beaulac, P.; Issa, M.; Ilinca, A.; Brousseau, J. Parameters Affecting Dust Collector Efficiency for Pneumatic Conveying: A Review. *Energies* **2022**, *15*, 916. [CrossRef]
4. Dudić, S.; Reljić, V.; Šešlija, D.; Dakić, N.; Blagojević, V. Improving Energy Efficiency of Flexible Pneumatic Systems. *Energies* **2021**, *14*, 1819. [CrossRef]
5. Ji, Y.; Liu, S.; Jia, J.; Yao, J. Analysis of the pick-up of crushed stone particles in a horizontal pipeline. *Powder Technol.* **2021**, *377*, 534–544. [CrossRef]
6. Ji, Y.; Hao, Y.; Yi, N.; Guan, T.; Gao, D. Particle flow regime in a swirling pneumatic conveying system. *Powder Technol.* **2022**, *401*, 117328. [CrossRef]
7. Marcus, R.D.; Leung, L.S.; Klinzing, G.E.; Rizk, F. *Pneumatic Conveying of Solids*; Springer: Dordrecht, The Netherlands, 2010.
8. Soepyan, F.B.; Cremaschi, S.; Sarica, C.; Subramani, H.J.; Gao, H. A model ranking and uncertainty propagation approach for improving confidence in solids transport model predictions. *J. Petrol. Sci. Eng.* **2017**, *151*, 128–142. [CrossRef]
9. Rice, H.P.; Fairweather, M.; Peakall, J.; Hunter, T.N.; Mahmoud, B.; Biggs, S.R. Constraints on the functional form of the critical deposition velocity in solid–liquid pipe flow at low solid volume fractions. *Chem. Eng. Sci.* **2015**, *126*, 759–770. [CrossRef]
10. Gomes, L.M.; Mesquita, A.L.A. Effect of particle size and sphericity on the pickup velocity in horizontal pneumatic conveying. *Chem. Eng. Sci.* **2013**, *104*, 780–789. [CrossRef]
11. Dasani, D.; Cyrus, C.; Scanlon, K.; Du, R.; Rupp, K.; Henthorn, K.H. Effect of particle and fluid properties on the pickup velocity of fine particles. *Powder Technol.* **2009**, *196*, 237–240. [CrossRef]
12. Kalman, H.; Satran, A.; Meir, D.; Rabinovich, E. Pickup (critical) velocity of particles. *Powder Technol.* **2005**, *160*, 103–113. [CrossRef]
13. Zhou, J.; Xu, L.; Du, C. Prediction of lump coal particle pickup velocity in pneumatic conveying. *Powder Technol.* **2019**, *343*, 599–606. [CrossRef]
14. Zhou, J.; Du, C.; Ma, Z. Influence of swirling intensity on lump coal particle pickup velocity in pneumatic conveying. *Powder Technol.* **2018**, *339*, 470–478. [CrossRef]
15. Zhou, J.; Du, C.; Liu, S.; Liu, Y. Comparison of three types of swirling generators in coarse particle pneumatic conveying using CFD-DEM simulation. *Powder Technol.* **2016**, *301*, 1309–1320. [CrossRef]
16. Ibrahim, K.A.; Hamed, M.H.; El-Askary, W.A.; El-Behery, S.M. Swirling gas-solid flow through pneumatic conveying dryer. *Powder Technol.* **2013**, *235*, 500–515. [CrossRef]
17. Yao, S.; Fang, T. Analytical solutions of laminar swirl decay in a straight pipe. *Commun. Nonlinear Sci.* **2012**, *17*, 3235–3246. [CrossRef]
18. Fokeer, S.; Lowndes, I.S.; Hargreaves, D.M. Numerical modelling of swirl flow induced by a three-lobed helical pipe. *Chem. Eng. Processing Process Intensif.* **2010**, *49*, 536–546. [CrossRef]
19. Fokeer, S.; Lowndes, I.; Kingman, S. An experimental investigation of pneumatic swirl flow induced by a three lobed helical pipe. *Int. J. Heat Fluid Fl.* **2009**, *30*, 369–379. [CrossRef]
20. Hui, L.I.; Tomita, Y. Measurements of Particle Velocity and Concentration for Dilute Swirling Gas-Solid Flow in a Vertical Pipe. *Particul. Sci. Technol.* **2002**, *20*, 1–13.
21. Li, H.; Tomita, Y. A Numerical Simulation of Swirling Flow Pneumatic Conveying in a Vertical Pipeline. *Particul. Sci. Technol.* **2001**, *19*, 355–368. [CrossRef]
22. Li, H.; Tomita, Y. Particle velocity and concentration characteristics in a horizontal dilute swirling flow pneumatic conveying. *Powder Technol.* **2000**, *107*, 144–152. [CrossRef]
23. Li, H.; Tomita, Y. An Experimental Study of Swirling Flow Pneumatic Conveying System in a Vertical Pipeline. *J. Fluids Eng.* **1998**, *120*, 200–203. [CrossRef]
24. Li, H.; Tomita, Y. An Experimental Study of Swirling Flow Pneumatic Conveying System in a Horizontal Pipeline. *J. Fluids Eng.* **1996**, *118*, 526–530. [CrossRef]
25. Zhou, L.X.; Chen, T. Simulation of swirling gas–particle flows using USM and $k - \varepsilon - k_p$ two-phase turbulence models. *Powder Technol.* **2001**, *114*, 1–11. [CrossRef]
26. Hayden, K.S.; Park, K.; Curtis, J.S. Effect of particle characteristics on particle pickup velocity. *Powder Technol.* **2003**, *131*, 7–14. [CrossRef]
27. Rabinovich, E.; Kalman, H. Incipient motion of individual particles in horizontal particle-fluid systems: A. Experimental analysis. *Powder Technol.* **2009**, *192*, 318–325. [CrossRef]

Article

Study on the Mechanism of Water Blocking in Tight Sandstone Gas Reservoirs Based on Centrifugal and Nuclear Magnetic Resonance Methods

Jianye Zhang [1,2], Yong Tang [1,*], Hongfeng Wang [2], Lan Huang [2], Faming Liao [2], Yongbing Liu [1] and Yiming Chen [3,*]

1   Faculty of Petroleum Engineering, Southwest Petroleum University, Chengdu 610500, China
2   Tarim Oilfield Branch, China National Petroleum Corporation, Korla 841000, China
3   Department of Petroleum Systems Engineering, University of Regina, Regina, SK S4S 0A2, Canada
*   Correspondence: tangyong2004@126.com (Y.T.); ycj293@uregina.ca (Y.C.)

Abstract: Tight sandstone gas reservoirs are characterized by deep burial, high pressure, tight matrix, fracture development, and the prevalence of edge and bottom water. Because of the small pore throats, the phenomenon of capillary force is evident. In addition, the low permeability of the reservoir and the difference in fluid properties make the gas reservoir undergo severe water-blocking damage. In this paper, centrifugal and nuclear magnetic resonance methods are used. The relationship between pore throat characteristics, fluid distribution, and gas reservoir water-blocking mechanism is studied and analyzed. The experimental results show that fracture formation increases the porosity of the small pores and expands the pore size distribution. It is conducive to the displacement of the bound water in the small pore space and the reduction in the bound water saturation. When increasing the same displacement pressure, the core porosity increases. More residual water in the tiny pores is converted to moveable water, thereby reducing the capillary resistance. The high-angle penetration fractures and complex seam networks are created by fractures. They connect the pores to form a fracture network structure, which is conducive to the communication of seepage channels. The increase in porosity and the creation of a complex fracture network make the water inrush along the fractures more even in the process of mining. This slows the advance speed of the water invasion front and reduces the damage of water blocking. The results enhance the understanding of the water invasion mechanism of edge and bottom water so as to improve the recovery factors of tight sandstone gas reservoirs.

Keywords: tight sandstone gas reservoir; nuclear magnetic resonance; fracture; water blocking

Citation: Zhang, J.; Tang, Y.; Wang, H.; Huang, L.; Liao, F.; Liu, Y.; Chen, Y. Study on the Mechanism of Water Blocking in Tight Sandstone Gas Reservoirs Based on Centrifugal and Nuclear Magnetic Resonance Methods. *Energies* **2022**, *15*, 6680. https://doi.org/10.3390/en15186680

Academic Editors: Gan Feng, Qingxiang Meng, Gan Li and Fei Wu

Received: 9 August 2022
Accepted: 6 September 2022
Published: 13 September 2022

Publisher's Note: MDPI stays neutral with regard to jurisdictional claims in published maps and institutional affiliations.

1. Introduction

As one of the most mature sources of unconventional gas resources, tight sandstone gas has very substantial reserves to be exploited. According to the research, the technically exploitable tight sandstone gas reserves today are 1050 to 2400 billion m³ [1]. However, tight sandstone gas reservoirs are characterized by deep burial, high pressure, tight matrix, fracture development, and the prevalence of edge and bottom water. Because of the small pore throats, the phenomenon of capillary force is obvious, and the difference in reservoir densities and fluid properties leads to severe water-blocking damage.

Water-blocking damage mainly refers to the phenomenon where the water saturation in the near-well zone of tight sandstone gas increases sharply because of the intrusion of foreign fluids into the reservoir, thus causing a decrease in gas-phase permeability. Once water blocking occurs in tight sandstone gas reservoirs, the permeability damage rate can reach more than 70%, and gas well production can drop to less than one-third of the original production [2]. Therefore, understanding the water-blocking mechanism of tight sandstone gas reservoirs is helpful to reduce water-blocking damage and improve production from gas reservoirs.

According to the different foreign fluids, Xu, S. Y. [3] divided the water-blocking damage process into two parts, capillary self-absorption and water phase retention, where capillary self-absorption refers to the rapid intrusion of edge and bottom water along the fracture and water phase retention refers to the inability to effectively return drainage because of the intrusion of working fluid into the reservoir. Wu, Y. [4] focused on the influence of water phase retention on water blocking. The study concluded that the longer the working fluid intrudes into the reservoir, the greater the water-bearing saturation and the more severe the permeability damage. Gao, Q. S. [5] clarified the effect of working fluid intrusion time on water-blocking damage by conducting damage experiments on tight sandstone reservoirs and classified the types of water blocking formed by working fluid into three types: deep intrusion low injury type, concave and convex intrusion medium injury type, and shallow intrusion high injury type. Using the pressure transfer test methods, Wang and Zhou [6] found that working fluid intrusion makes the sandstone permeability decrease, which aggravates the water-blocking damage, and finally proposed a nanoemulsion system that effectively relieves the water-blocking damage. Zhang et al. [7] proposed that when the surface tension of the working fluid is large, the reservoir core is severely damaged by water blocking, and foreign fluids are not easily displaced. Qin et al. [8] investigated the effect of the fracture morphology and developmental characteristics on water blocking by using SEM and CT scans and found that intragrain fractures connect the pores and form a seepage network in dense sandstones, which greatly increases the permeability. Dong et al. [9] divided the fractures in the Kalasu tight gas reservoir into macroscopic fractures and microscopic fractures based on the fracture pore size, fracture network effectively improves water-blocking damage by increasing permeability and gas seepage performance. Scholars at home and abroad have mainly studied the influence of working fluid on water locking in tight sandstone gas reservoirs while ignoring the study of the mechanism of edge and bottom water on water blocking in gas reservoirs.

As an emerging high technology, nuclear magnetic resonance (NMR)scanning technology is widely used in many aspects of oil and gas exploration. However, the current NMR scanning technique is less applied to the interpretation of the water-blocking mechanism and is mainly used to study the occurrence state and mobility of pore water in tight gas reservoirs. Yao et al. [10] combined gas–water two-phase flow and NMR and found that whenever the saturation of the water-phase fluid entering the reservoir exceeded the original saturation, it would cause reservoir water-blocking damage. Li et al. [11] used NMR testing to quantify the extent of water-blocking damage from a microscopic perspective and concluded that the root of relieving water-blocking damage is to eliminate the occupation of storage space by capillary water. Li et al. [12] integrated NMR testing and core flooding, and the results showed that the lesser the percentage of movable water in the pore space, the more serious the degree of water blocking in the reservoir core under saturated water conditions. Zhang et al. [13], using the NMR technique, analyzed the mechanism of water-blocking damage in working fluids from a microscopic perspective. Hassan et al. [14] used NMR measurements to find that injecting a thermochemical fluid into a dense sample can reduce capillary forces and undo water-blocking damage.

In summary, the existing water-blocking mechanism of tight sandstone gas reservoirs mainly focuses on the water-phase retention caused by the working fluid, ignoring the study of the water-blocking mechanism of edge and bottom water. While NMR is an effective tool for oil and gas detection, it is less applied to explain the water-blocking mechanism in gas reservoirs. Therefore, this study comprehensively considered the impact of parameters such as porosity, water saturation, displacement pressure, pore size distribution, and fracture type on water blocking. Firstly, the core data of six wells in a tight sandstone gas reservoir were selected, and then centrifugal and NMR methods were used to conduct gas-displacing water experiments before and after fracture formation to simulate the effect of fracture formation on the water block in the reservoir. Finally, the water-blocking mechanism of gas reservoirs was analyzed based on the results of gas-flooding efficiency, expanding our understanding of the water-blocking mechanism in tight sandstone gas reservoirs.

2. Experimental Method

2.1. Gas Displacing Water Experiment

The centrifugal method uses the centrifugal force of a high-speed centrifuge instead of the displacing pressure. This study mainly used the difference in centrifugal force generated by the difference in density between the nonwetting phase and the wetting phase to overcome the capillary force to drive out the water in the tiny pores and finally achieve the effect of displacement. The NMR scan analysis can analyze the fluid distribution characteristics in the pore space based on the measured T_2 spectrum curve. When the core pore throat is completely saturated with water, the larger the pore space is, the longer the T_2 relaxation time, and vice versa.

Therefore, gas-displacing water experiments were performed before and after fracture formation to simulate its impact on the gas reservoirs. Firstly, based on the NMR T_2 spectrum, the occurrence (movable or bound) state of brine within the core pores was analyzed. Then, according to the analysis results, the movable fluid saturation and the bound fluid saturation were given quantitatively, which helped us study the effect of the gas-driving efficiency on the gas reservoir. Finally, the water-blocking mechanism of tight sandstone gas reservoirs was clarified by exploring the influence of different factors on water-blocking damage.

2.2. Experimental Steps of Gas Displacing Water

The matrix plunger core was first evacuated. Secondly, an NMR scan was performed after saturation with water, in which its echo interval was 0.2 ms. Then centrifugation was performed at different speeds of 1500/3000/5000/7000/9000/10,500/12,000 turn/min, and an NMR scan with an echo interval of 0.2 ms was performed after each centrifugation. Therefore, the capillary pressure curve could be calculated using each NMR scan. Immediately afterward, according to the experimental method of matrix cores, the plunger cores were fractured, and centrifugal and NMR were performed to obtain their capillary pressure curves after fractures. Finally, by processing the experimental data, the distribution characteristics of core pore throats in a tight sandstone gas reservoir were obtained, and the water-blocking mechanism was analyzed. Table 1 shows the experimental conditions for NMR testing.

Table 1. Experimental conditions for NMR testing.

Parameter	Value
Testing temperature/°C	35
Waiting time/s	6
Echo interval/ms	0.20
Number of scanning	128
Acceptance gain/%	80
Number of echoes	4096

2.3. Core Basic Physical Properties

A typical six-well core from a tight sandstone gas reservoir was selected for porosity and permeability measurements. Core porosity was measured by mercury intrusion porosity, and permeability was measured by the transient pulse method. Finally, the following core basic physical parameters were obtained (Table 2).

Table 2. Basic physical parameters of cores.

Number	Permeability/mD	Porosity/%
No. 12 of Well XX10	0.098	5.12
No. 20 of Well XX208	0.191	6.40
No. 35 of Well XX2-2-5	0.033	6.10
No. 94 of Well XX2-1-5	0.019	4.20
No. 7 of Well XX2-1-5	0.041	5.60
No. 21 of Well XX1003	0.107	7.07

3. Experimental Results

To study the effects of porosity, water saturation, displacement pressure, pore size distribution, and fracture morphology on water blocking in gas reservoirs, centrifugal and NMR experiments were performed. Finally, by comparing these effects on the gas-driving efficiency before and after fracture formation, the characteristics of pore throat distribution in tight sandstone gas reservoirs were obtained, and the effects on reservoir water blocking were analyzed.

3.1. Effect of Changes in Porosity and Water Saturation on Water Blocking after Fracture

The cores from the tight sandstone gas reservoir No. 64 of well XX1003 were selected, and their plunger cores were tested by centrifugation and NMR. We finally obtained the NMR T_2 spectrum curves and core pore size distributions at different rotational speeds before and after fracture, where the rotational speeds were 1500/3000/5000/7000/9000/10,500/12,000 turn/min, respectively. When the core is completely saturated with water, the T_2 spectrum curve is the uppermost peripheral one, and its vertical coordinate is the porosity component of all filled water.

Comparing the NMR T_2 spectrum curves of cores before and after fracture formation at different rotational speeds (Figure 1), it is found that, although the T_2 spectrum curves before and after fracture were both bimodal, the right peak of the curve before fracture was smaller and the left peak (short T_2 relaxation time) was dominant. It indicates that the core before fracture is mainly small pores, in which the water is poorly movable, and the capillary resistance is high, so the phenomenon of water blocking is serious. The right peak (T_2 relaxation time) of the curve is longer after fracture, which indicates that there are more large pores and stronger water movability. Therefore, the difference between the capillary force of the water phase and gas phase is smaller, the capillary resistance is reduced, and finally, the water-blocking phenomenon is effectively relieved.

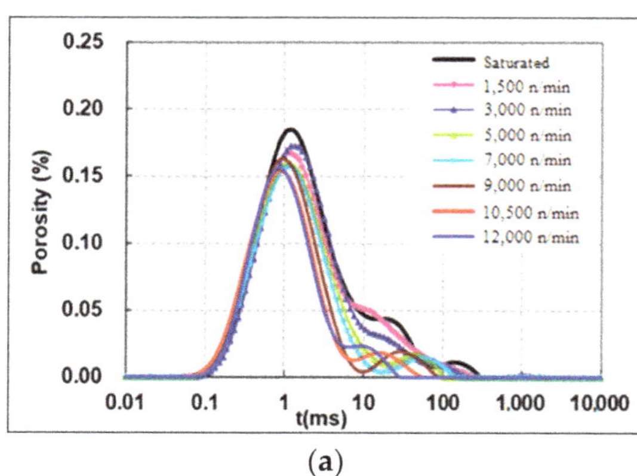

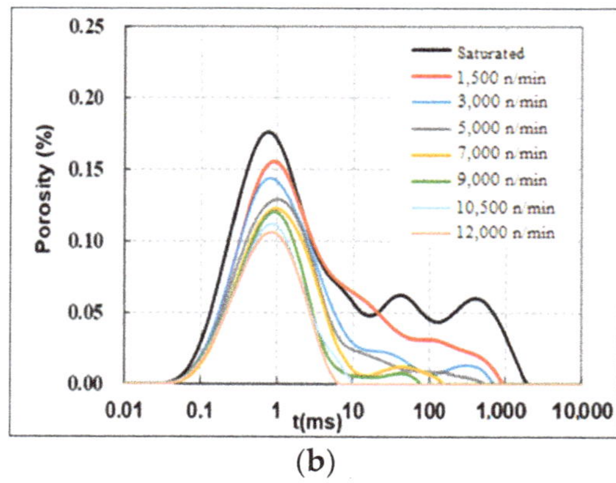

(**a**)　　　　　(**b**)

Figure 1. NMR T_2 spectrum curves at different rotational speeds: (**a**) before fracture; (**b**) after fracture.

From the curve characteristics of the core before and after fracture, the porosity of the core after fracture is relatively larger, so the water in the pore space is gradually driven out with the continuous gas repulsion. The T_2 spectrum curve after gas flooding is shifted more, the final residual water saturation is lower, and finally, the degree of water-blocking damage is gradually reduced. In summary, the fracture is conducive to increasing the porosity of tight gas reservoirs and reducing capillary resistance. As the gas continues to occupy the pore space, the bound water keeps flowing out from the pore, which reduces the residual water saturation while slowing down the capillary self-absorption effect and finally effectively releasing the water-blocking damage.

Further, the distribution characteristics of gas and water in the pore space before and after fracture can be seen from the changes in water saturation at different speeds before and after fracture (Table 3). Before the fracture, with the increase in centrifugal speed, the centrifugal force gradually overcomes the capillary force. The gas gradually occupies the space in the pore, which makes the water saturation decrease. However, the large pore content is low and poorly connected, and the permeability devotion value is dominated by the large pore. Although the pressure is sufficient, it is difficult to completely block the water phase. Therefore, even if the speed reaches 1200 turn/min, there is still about 70% of the water that cannot be effectively detached, and the saturation of bound water by centrifugation is high. It indicates that the sample is very dense and that a large amount of water has not been displaced. Even if the centrifugal speed is increased, the effect of gas-driving water is limited, and the water lock phenomenon is serious. The porosity of the core increases significantly after the fracture is created. Since the gas and water flow preferentially in the large pore space, the water saturation in the large pore space decreases rapidly, and the water is gradually driven out. However, because of the increase in the porosity in small pores, the effective seepage channels of the reservoir increase, and the water in the small pores is gradually replaced by gas. Therefore, the water saturation is continuously decreasing, and only about one-third of the water in the dense micropores has not been effectively displaced, which shows that the fracture-making effect is significant. When the speed reaches 12,000 turn/min, the saturation of bound water is only 34%. Therefore, the fracture structure can effectively communicate with the seepage channels and increase the porosity of the small pores while displacing the bound water from the small pores. Eventually, the gas permeability is increased by reducing the saturation of bound water to achieve effective improvement of water-blocking damage.

Table 3. Changes of water saturation at different rotational speeds before and after fracture.

Rotational Speed (Turn/min)	Saturation Measured by NMR before Fracture/%	Saturation Measured by NMR after Fracture/%
0	100.00	100.00
1500	94.65	79.03
3000	90.36	61.72
5000	81.73	53.58
7000	75.94	46.68
9000	73.89	40.13
10,500	69.31	36.91
12,000	66.67	34.05

3.2. Effect of Changes in Displacement Pressure Curve on Water Blocking after Fracture

The graph of the displacement pressure before fracture is shown in Figure 2 and indicates that when the displacement pressure increases to 0.7 MPa, only less than 20% of the water is displaced by the gas. Because this part of water exists in the large pores of the core, it can be displaced with only a small pressure. When the displacement pressure increased from 0.7 to 3 MPa, the gas flow rate increased because the displacement pressure overcomes part of the capillary force and drives part of the water flow in the pore space, eventually displacing about 10% of the water. When the displacement pressure increased to 4 MPa, the presence of a dense core and small porosity made the seepage channel resistance higher, so the water saturation of the core was around 70%, and still, more than two-thirds of the water was not displaced. This indicates that the gas flooding effect is still poor, and the water-blocking effect occurs in the reservoir, which leads to a significant reduction in gas production and ultimately affects the production of gas wells.

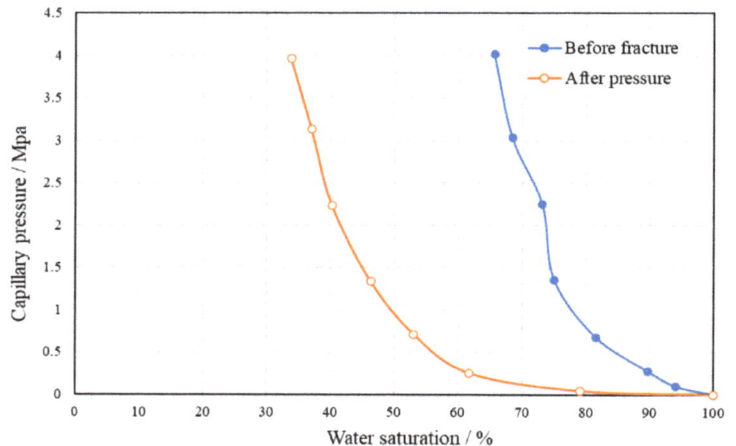

Figure 2. Variation of displacement pressure.

Fracture makes the core porosity increase, and more residual water in the tiny pores is transformed into movable water, which effectively communicates the seepage channels in the core. Therefore, when the displacement pressure is increased to 0.7 MPa, more than 45% of the water is displaced by the gas, compared with less than 20% of the water before fracture. When the displacement pressure is increased from 0.7 to 3 MPa, with the increase in displacement pressure, the capillary force of tiny pores is overcome. Therefore, more water in the pores and channels becomes movable water to be driven out. It was found that about 15% of water is driven out, and the relative permeability of the gas phase is increased. When the displacement pressure reached 4 MPa, only about one-third of the water remained because of the existence of even smaller pores and higher capillary resistance. Compared with the effect of displacement before fracture, although there are some tiny pores after fracture formation, fractures improve the overall physical properties of the reservoir and makes the core porosity increase significantly, which effectively communicates the seepage channel. At the same strength, the displacement pressure overcomes the capillary force of most of the tiny pores, further driving out the remaining water and alleviating the water-lock effect.

3.3. Effect of Changes in Pore Size Distribution on Water Blocking after Fracture

Centrifugal and NMR tests were carried out on the plunger cores from six wells in a tight sandstone gas reservoir to obtain pore size distribution maps before and after fracture, of which five wells were analyzed primarily (Figures 3–7).

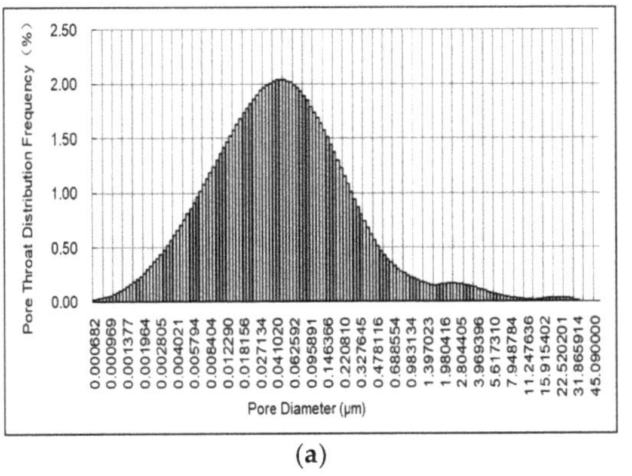
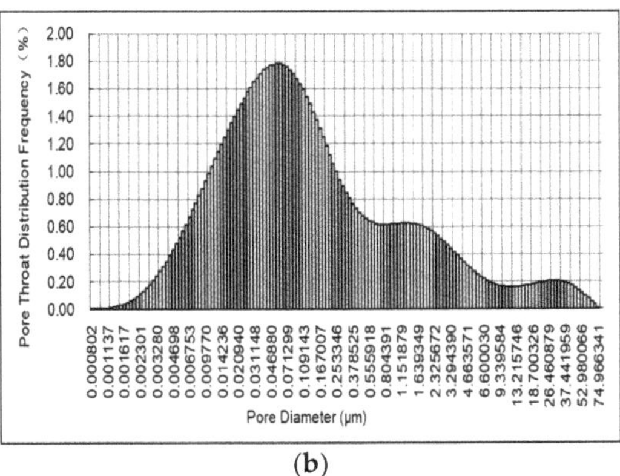

(**a**)

(**b**)

Figure 3. Distribution of core pore size (No. 21 of Well XX1103): (**a**) before fracture; (**b**) After fracture.

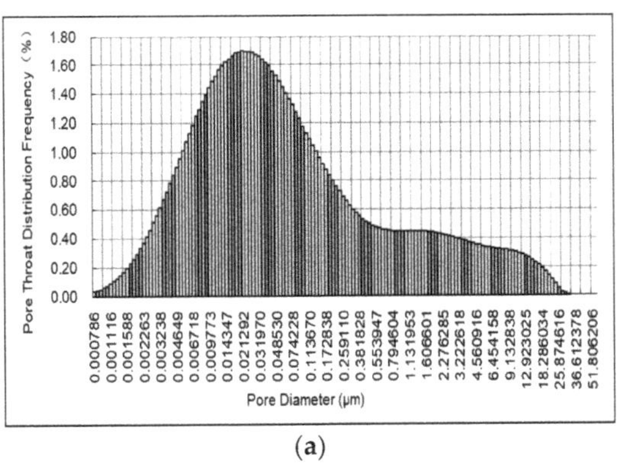
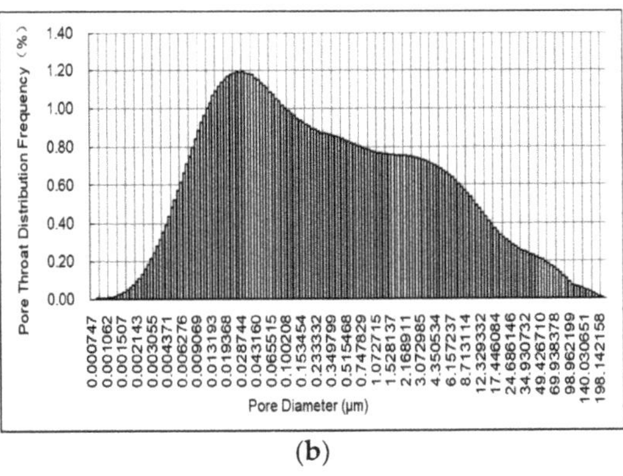

(**a**)

(**b**)

Figure 4. Distribution of core pore size (No. 3 of Well XX2-2-3): (**a**) before fracture; (**b**) After fracture.

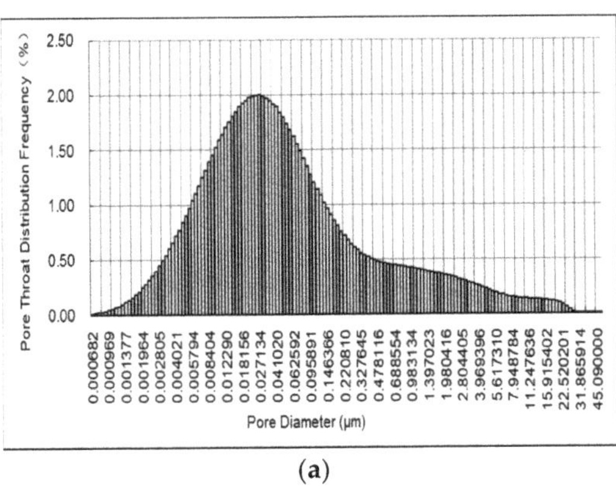
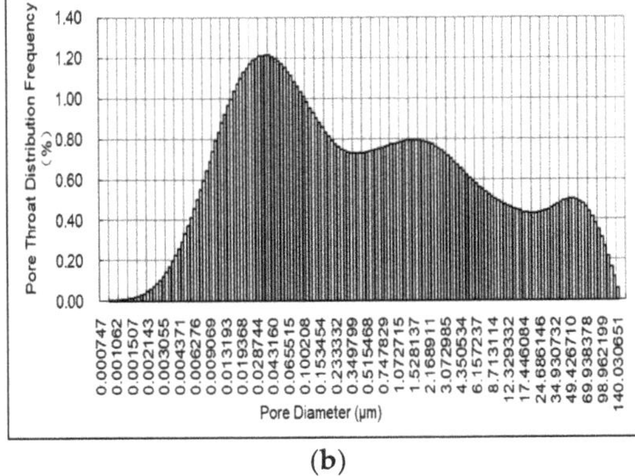

(**a**)

(**b**)

Figure 5. Distribution of core pore size (No. 9 of Well XX2-1-5): (**a**) before fracture; (**b**) After fracture.

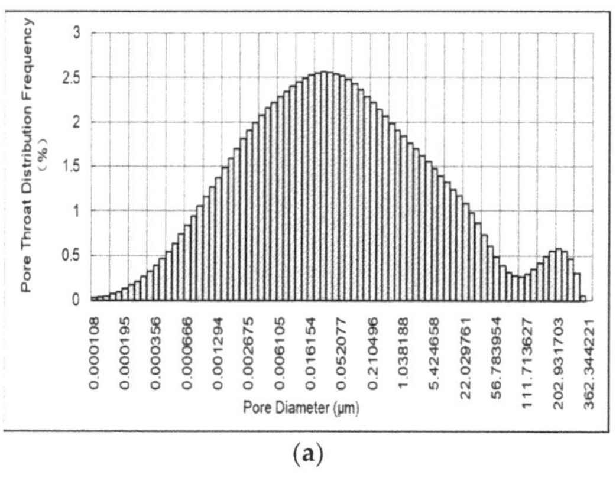

(a)

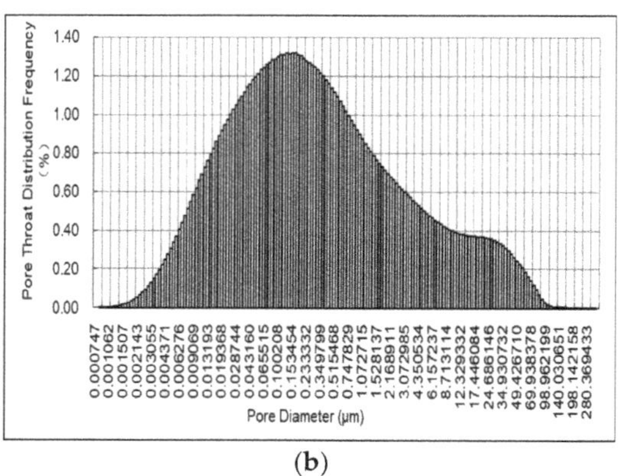

(b)

Figure 6. Distribution of core pore size (No. 19 of Well XX1003): (**a**) before fracture; (**b**) After fracture.

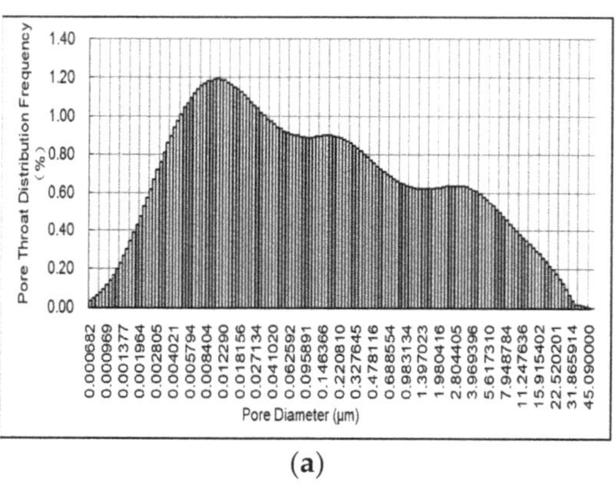

(a)

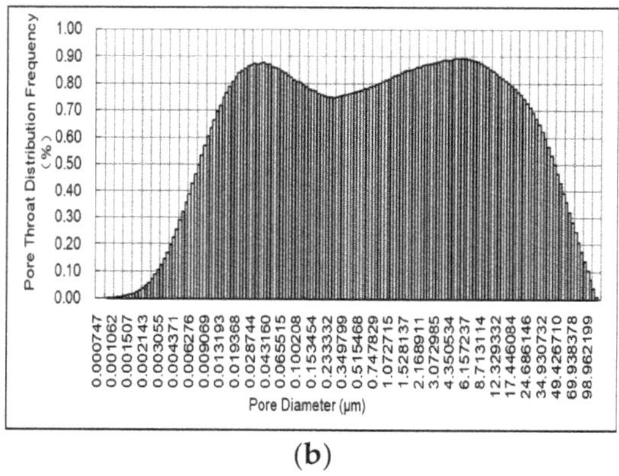

(b)

Figure 7. Distribution of core pore size (No. 51 of Well XX2-1-5): (**a**) before fracture; (**b**) after fracture.

According to the basic shape of the pore size distribution before and after fracture, it is found that the pore size distribution of the core before fracture is mainly unimodal and bimodal, and the main peak is concentrated on the left side, while the side peaks are not obvious. This feature indicates that the overall pore size before fracture is very small, with more than half of the pore size less than 20 nm, which makes the pore type complex, resulting in poor connectivity between large and small pores and high capillary resistance. Therefore, during the displacement process, it mainly shows the fluid output in the large pore space and the difficulty of fluid flow in the small pore space, resulting in poor gas-driving effect and serious water-blocking phenomenon. After a fracture, it is observed that both the primary and secondary peaks showed a tendency to move to the right, and the secondary peaks became larger. It means that the proportions of the large pore size become larger after fracture, and the proportions of small pore size become smaller, while the ratio of pore sizes less than 20 nm is only about one-third. The pore size distribution becomes wider, and the connectivity between large and small pores is well improved, which weakens the nonhomogeneity of the reservoir. During the displacement process, as the capillary resistance of the fluid decreases, more water in the pore space is displaced by the gas, and the residual water saturation decreases, which ultimately improves the water-blocking phenomenon effectively.

Comparing the average values of pore size distribution before and after fracture (Figure 8), it is found that the center of the core pore size distribution has shifted towards large sizes. From the overall view of the five samples, it is found that the ratio of pore throat of less than 20 nm decreased by 19.82%, the ratio of 20 nm–100 nm decreased by 0.25%, the ratio of 100 nm–1 μm increased by 6.63%, and the ratio of greater than 1 μm increased by 13.44% after the fracture. Fractures widen the pore size distribution, thereby improving fluid flow channels and converting bound water in small pores into moveable water. By reducing the capillary resistance, the bound water in the small pore space is displaced out, making the residual water saturation reduced and mitigating the water-blocking effect.

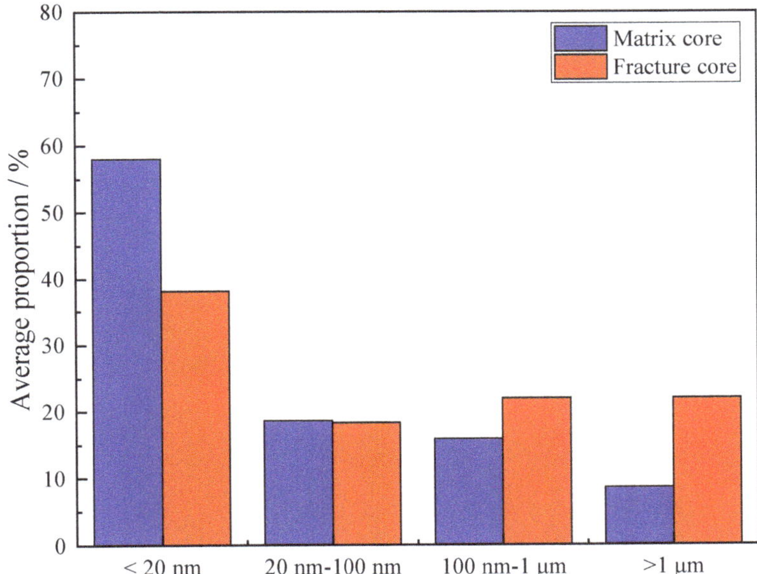

Figure 8. Comparison of the average value of pore size distribution before and after a fracture.

3.4. Effect of Fracture Morphology on Water Blocking after Fracture

Fractures that can penetrate through the sandstone level are called penetrating fractures [15]. Based on the relationship between the direction of the wellbore and the direction of the minimum horizontal principal stress, the fracture morphology can be divided into three types: lateral penetrating fracture, high-angle penetrating fracture, and complex fracture networks [16]. Among them, the lateral penetrating fracture produces a fracture surface parallel to the ground, the high-angle penetrating fracture produces a fracture surface perpendicular to the ground, and the complex fracture networks produce a fracture surface that is both perpendicular and parallel to the ground.

Observing the fracture morphology diagram after a fracture (Figure 9) and the relationship between fracture morphology and gas-driving efficiency before and after fracture creation (Table 4), it is found that the fracture morphology in No. 3 of well XX2-2-3 after fracture morphology is a lateral penetrating fracture. Its fracture porosity only increased by 0.34%, and the gas-driving efficiency was poor, increasing by only 28.10%. The fracture morphology of No. 64 in Well XX1003 is a high-angle penetrating fracture. Although the fracture porosity only increased by 1.85% after fracture, its gas-driving efficiency increased by 97.87%, which significantly relieves the water-blocking damage. The fracture morphology of No. 21 in Well XX1103 is a complex fracture network. The fracture has increased its fracture porosity by 1.99%, significantly improved the gas-driving efficiency by 95.86%, and reduced the water-blocking effect.

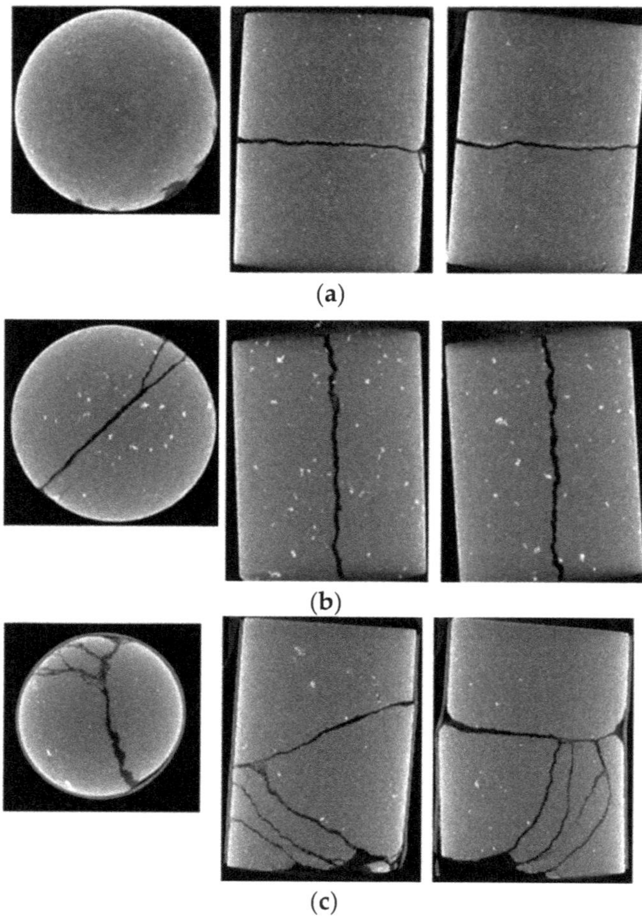

Figure 9. Fracture morphology after fracture: (a) No. 3 of Well XX2-2-3; (b) No. 64 of Well XX1003; (c) No. 21 of Well XX1003.

Table 4. Relationship between fracture morphology and gas-driving efficiency before and after fracture in three wells.

Core Number	Fracture Morphology	Matrix Samples		Fracture Samples		Fracture Porosity	Gas-Driving Efficiency Improvement
		NMR Porosity	Gas-Driving Efficiency	NMR Porosity	Gas-Driving Efficiency		
No. 3 in Well XX2-2-3	Lateral penetrating fracture	6.95%	46.27%	7.29%	59.27%	0.34%	28.10%
No. 64 in Well XX1003	High-angle penetrating fracture	4.91%	33.33%	6.76%	65.95%	1.85%	97.87%
No. 21 in Well XX1003	Complex fracture networks	6.19%	29.26%	8.18%	57.31%	1.99%	95.86%

In comparison, high-angle penetrating fracture and complex fracture networks are found to be more effective for enhancing gas and water displacement. This is because these two types of fractures have a large number of longitudinal fractures near the end of the production layer and a larger width of the fracture body, which effectively communicates the seepage channel. This increases the seepage area of the gas well while reducing the capillary resistance of the gas in the fracture, which ultimately helps to reduce the water-blocking effect. However, lateral penetrating fracture produces fractures parallel to the ground, and the width of the fracture body is small, which cannot effectively communicate the gas seepage channel. Therefore, the lateral penetrating fracture is less effective in enhancing the efficiency of gas-driving water and cannot effectively relieve the water-blocking damage.

3.5. Effect of Fracture Making on Water Blocking of Edge and Bottom

Tight sandstone gas reservoirs are characterized by fracture development and edge–bottom water. In the process of gas reservoir development, the edge–bottom water first bursts rapidly along the fracture and advances in a piston-like manner in the matrix. The occurrence of capillary fingering and snap-off causes reservoir water-blocking damage. Then, the production pressure difference makes the bottom water quickly channel to the local gas well along the fracture. The higher the production pressure difference, the faster the water runs. As a result, many gas wells produce water within a short period of time [17–21].

After fracture formation, centrifugal and NMR experiments show that the fracture morphology is characterized by high-angle penetrating fracture and complex fracture networks. Through the connected pores, the fracture network structure is formed. This not only increases the effective seepage channels but also reduces the capillary force. At the same time, the porosity of small pores and the proportion of large pore sizes are increased. Therefore, the bound water of small pore size is converted into movable water, and the proportion of movable water is increased. With the significant increase in matrix permeability, the ability to supply gas to the fractures is greatly enhanced, further improving the heterogeneity of the gas reservoir. As the displacement pressure increases, the bound water in the small pore space is displaced out by the gas, and the saturation of bound water is decreased accordingly. This can effectively improve the water-blocking damage. At this time, the edge and bottom water advance uniformly in the matrix. When the fracture is connected to the water body, the water body is more uniformly protruding along the fracture during the mining process. The advance speed of the water invasion front is slowed, and the time to water appearance is delayed, which significantly improves gas reservoir recovery.

4. Conclusions

Using centrifugal and NMR methods, gas displacing water experiments were conducted on six well cores from tight sandstone gas reservoirs before and after fracture, and the following conclusions were finally obtained:

(1) The porosity of small pores is increased, and the capillary resistance is reduced. Therefore, the bound water can be continuously displaced from pores, and the water saturation decreases, which can effectively reduce the water-blocking damage;

(2) When the same displacement pressure is increased, the porosity of the core after fracture increases significantly. This makes more residual water in the tiny pores to be transformed into movable water, which effectively communicates the seepage channels and reduces the water-blocking damage;

(3) After the fracture, the pore size distribution becomes wider, and the proportion of large pore size increases. As the bound water in the small pore space is transformed into movable water, the proportion of movable water increases, and the residual water saturation decreases, which can effectively improve the water-blocking phenomenon;

(4) After the fracture, high-angle penetrating fracture and complex fracture networks are more likely to produce larger width of the fracture body than lateral penetrating fracture. This facilitates the communication of seepage channels and significantly reduces the water-blocking effect while reducing the capillary resistance of the gas in the fracture;

(5) The porosity of the pores is increased by forming fractures. When the fractures are connected to the pore space, the fracture network structure is formed. Therefore, edge–bottom water is more evenly protruding along the fractures in the process of

extraction, which slows down the advance speed of the water invasion front and improves the recovery rate of the gas reservoir.

Author Contributions: Conceptualization, Y.T., F.L. and Y.L.; data curation, Y.T., J.Z. and H.W.; funding acquisition, Y.T. and F.L.; investigation, J.Z., F.L. and Y.L.; methodology, H.W. and L.H.; project administration, J.Z.; writing—original draft, Y.T.; writing—review & editing, Y.T. and Y.C. All authors have read and agreed to the published version of the manuscript.

Funding: This research received no external funding.

Data Availability Statement: The data presented in this study are available on request from the corresponding author.

Conflicts of Interest: The authors declare no conflict of interest.

References

1. You, L. The Study of Mechanisms of Aqeous Phase Trapping and its Applications in Tight Sands Gas Reservoirs. Master's Thesis, Southwest Petroleum University, Chengdu, China, 2004.
2. Su, Y.; Zhuang, X.; Li, L.; Li, X.; Li, D.; Zheng, J. Experiment on Long Core Pressure Loss and Permeability Damage Evaluation in Tight Sandstone Gas Reservoir. *Spec. Oil Gas Reserv.* **2021**, *28*, 98–102.
3. Xu, S. Evaluation of Water Blocking Damage in Ultra-tight Sandstone Gas Reservoirs. Master's Thesis, Southwest Petroleum University, Chengdu, China, 2017.
4. Wu, Y. The Experimental Research of Water-Blocking Damage and Removal Methods for the Ordos Tight Sandstone Gas Reservoirs. Master's Thesis, China University of Petroleum, Beijing, China, 2018.
5. Gao, Q. Analysis of the Damage Characteristics of the Tight Sandstone Gas Reservoir Water Lock in Daniudi Gas Field. *Sci. Technol. Eng.* **2018**, *18*, 156–162.
6. Wang, J.; Zhou, F. Cause Analysis and Solutions of Water Blocking Damage in Cracked/non-cracked Tight Sandstone Gas Reservoirs. *Pet. Sci.* **2020**, *18*, 219–233. [CrossRef]
7. Zhang, J.; He, W.; Hao, T.; Wen, X. Experimental Evaluation of Damage Factors of Tight Sandstone Gas's Reservoirs in Shenfu Block, China. *Fresenius Environ. Bull.* **2020**, *29*, 6951–6959.
8. Qin, S.; Wang, R.; Shi, W.; Liu, K.; Zhang, W.; Xu, X.; Qi, R.; Yi, Z. Diverse Effects of Intragranular Fractures on Reservoir Properties, Diagenesis, and Gas Migration: Insight from Permian Tight Sandstone in the Hangjinqi Area, North Ordos Basin. *Mar. Pet. Geol.* **2022**, *137*, 105526–105543. [CrossRef]
9. Dong, Y.; Lu, X.; Fan, J.; Zhuo, Q. Fracture Characteristics and Their Influence on Gas Seepage in Tight Gas Reservoirs in the Kelasu Thrust Belt (Kuqa Depression, NW China). *Energies* **2018**, *11*, 2808. [CrossRef]
10. Yao, J.; Wang, H.; Pei, G.; Yuan, S.; Zhang, H. The Formation Mechanism of Upper Paleozoic Tight Sand Gas Reservoirs with Ultra-low Water Saturation in Eastern Ordos Basin. *Nat. Gas Ind.* **2014**, *34*, 37–43.
11. Li, N.; Wang, Y.; Zhang, S.; Li, J.; Zhang, Z.; Zhao, C.; Huang, W. Study on Water Block in Tight Sandstone Gas Reservoirs and Solutions Thereof. *Drill. Fluid Completion Fluid* **2016**, *33*, 14–19.
12. Li, X. Study on Damage and Cleanup of Water Blocks in Low Permeability Sandstone Reservoirs in SuDong Block. Master's Thesis, China University of Petroleum, Qingdao, China, 2019.
13. Zhang, L.; Zhou, F.; Zhang, S.; Li, Z.; Wang, J.; Wang, Y. Evaluation of Permeability Damage Caused by Drilling and Fracturing Fluids in Tight Low Permeability Sandstone Reservoirs. *J. Pet. Sci. Eng.* **2019**, *175*, 1122–1135.
14. Hassan, A.M.; Mahmoud, M.A.; Al-Majed, A.A.; Al-Nakhli, A.R.; Bataweel, M.A. Water Blockage Removal and Productivity Index Enhancement by Injecting Thermochemical Fluids in Tight Sandstone Formations. *J. Pet. Sci. Eng.* **2019**, *192*, 106298–106307. [CrossRef]
15. Zeng, Q.; Zhang, R.; Lu, W.; Wang, B.; Wang, C. Fracture Development Characteristics and Controlling Factors Based on 3D Laser Scanning Technology: An Outcrop Case Study of Suohu Village, Kuqa Foreland Area, Tarim Basin. *Nat. Gas Geosci.* **2017**, *28*, 397–409.
16. Tang, Y.; Li, J.; Zhang, H.; Zhu, Q.; Liu, S.; Zong, X. Study of Seepage Mechanisms in Fractured Horizontal Gas Wells. *China Pet. Chem. Stand. Qual.* **2012**, *32*, 62–103.
17. Sang, D. Research of Two-phase Flow Mechanism of Gas Water Interaction in Puguang Gas Field During the Water Invasion. Master's Thesis, Southwest Petroleum University, Chengdu, China, 2018.
18. Hu, Y.; Mei, Q.; Chen, Y.; Guo, C.; Xu, X.; Jiao, C.; Jia, Y. Study on Water invasion Mechanism and Control Technology of Edge and Bottom Water Gas Reservoirs. In Proceedings of the 31st Annual National Natural Gas Symposium (2019), Lisbon, Portugal, 10–12 July 2019; pp. 170–179.
19. Fang, F.; Shen, W.; Li, X.; Gao, S.; Liu, H.; Li, J. Experimental study on water invasion mechanism of fractured carbonate gas reservoirs in Longwangmiao Formation, Moxi block, Sichuan Basin. *Environ. Earth Sci.* **2019**, *78*, 316. [CrossRef]
20. Guo, C.-F.; Li, H.-B.; Tao, Y.; Lang, L.-Y.; Niu, Z.-X. Water invasion and remaining gas distribution in carbonate gas reservoirs using core displacement and NMR. *J. Cent. South Univ.* **2020**, *27*, 531–541. [CrossRef]
21. Zhou, M.; Li, X.; Hu, Y.; Xu, X.; Jiang, L.; Li, Y. Physical simulation experimental technology and mechanism of water invasion in fractured-porous gas reservoir: A review. *Energies* **2021**, *14*, 3918. [CrossRef]

Article

Study on the Accuracy of Fracture Criteria in Predicting Fracture Characteristics of Granite with Different Occurrence Depths

Chenbo Liu [1], Gan Feng [1,2,*], Hongqiang Xie [1,*], Jilan Wang [1], Zhipan Duan [1], Ye Tao [1], Gongda Lu [1], Huining Xu [1], Yaoqing Hu [3], Chun Li [3], Yuefei Hu [3], Qiuhong Wu [4] and Lu Chen [5]

[1] College of Water Resource & Hydropower, Sichuan University, Chengdu 610065, China
[2] Ministry of Education, Key Laboratory of Deep Earth Science and Engineering, Sichuan University, Chengdu 610065, China
[3] College of Mining Engineering, Taiyuan University of Technology, Taiyuan 030024, China
[4] Work Safety Key Lab on Prevention and Control of Gas and Roof Disasters for Southern Coal Mines, Hunan University of Science and Technology, Xiangtan 411201, China
[5] School of Civil Engineering, Changsha University of Science & Technology, Changsha 410000, China
* Correspondence: fenggan@whu.edu.cn (G.F.); alex_xhq@scu.edu.cn (H.X.);
  Tel.: +86-135-0971-7523 (G.F.); +86-172-0828-1067 (H.X.)

Abstract: The fracture network of a deep geothermal reservoir forms the place for heat exchange between injected fluid and rock mass with high temperature. The fracture resistance ability of reservoir rocks will affect the formation of fracture-network structure, heat exchange and transmission characteristics, and reservoir mechanical stability. However, there are few reports on the fracture toughness and trajectory prediction of geothermal reservoirs with different depths. In this paper, the modified maximum tangential stress criterion (MMTS) is analyzed. The results show that the experimental data are significantly different from the theoretical estimate of MMTS under the influence of different occurrence depths. It is found that the fracture process zone (FPZ) seriously affects the accuracy of predicting fracture initiation angle and mixed-mode (I+II) fracture toughness by MMTS. The FPZ value, considering the influence of different occurrence depths, is modified, and the accuracy of MMTS in predicting the fracture mechanical characteristics of granite is improved. In addition, the mechanical test results show that the Brazilian splitting strength (σ_t) of granite fluctuates increase with the increase in temperature. With the increase in deviatoric stress, the Brazilian splitting strength and the Brazilian splitting modulus of rock show a trend of first increasing, then decreasing, and then increasing.

Keywords: rock mechanical properties; MMTS; geothermal development; geothermal reservoir; different depth of occurrence

Citation: Liu, C.; Feng, G.; Xie, H.; Wang, J.; Duan, Z.; Tao, Y.; Lu, G.; Xu, H.; Hu, Y.; Li, C.; et al. Study on the Accuracy of Fracture Criteria in Predicting Fracture Characteristics of Granite with Different Occurrence Depths. *Energies* **2022**, *15*, 9248. https://doi.org/10.3390/en15239248

Academic Editor: José António Correia

Received: 22 October 2022
Accepted: 4 December 2022
Published: 6 December 2022

Publisher's Note: MDPI stays neutral with regard to jurisdictional claims in published maps and institutional affiliations.

1. Introduction

Geothermal energy has the advantages of being clean and renewable, having wide distribution and high heat storage, and is regarded as a renewable energy with great potential [1–4]. It can be used for power generation, heating, planting, breeding, and so on [5]. According to the heat storage temperature and burial depth, it can be divided into shallow geothermal resources and deep geothermal resources [6–8]. According to theoretical estimates, the total amount of geothermal energy stored within the upper 10 km of the Earth's crust is about 1.3×10^{27} J, which can be used for approximately 21.7 million years globally [9,10]. Deep geothermal resources are stored in hot dry rock. The hot dry rock mass is compact and has very little permeability. Traditional geothermal extraction methods are difficult to exploit effectively, so it is necessary to build an enhanced geothermal system (EGS). In recent years, the construction of EGS projects has been carried out. According to statistics, by the end of 2021, 41 EGS projects had been built across the world [11], such as the Fenton Hill Project in the United States [12], the Rosemanowes project in

Britain [13], the Mauerstetten project in Germany [14], and the Soultz project in France [15]. Among them, the most successful project is the Soultz project in France, which has a power generation capacity of megawatts [15]. The ability of the hot dry rock reservoir to resist fracture damage directly affects its fracture-network structure, heat transfer characteristics, and reservoir mechanical stability. Therefore, it is necessary to study the failure fracture mechanical properties of geothermal reservoirs. However, due to the complex geological environment of geothermal reservoirs and the different occurrence depths of reservoirs, there is a lack of research on the fracture mechanics behavior of geothermal reservoirs.

The fracture toughness of rock is an important element in fracture mechanics and reflects the ability of material to resist crack propagation. Previous studies have shown that modified maximum tangential stress (MMTS) theory is suitable for the fracture mechanics of brittle materials such as rocks, and it is commonly used to predict the fracture toughness and crack initiation angle of rocks. Ayatollahi et al. [16] compared the predicted value of fracture toughness ratio (K_{IIC}/K_{IC}) with the experimental value using the MTS criterion considering stress singularity, T-stress, and the second non-singular term, and the results showed that the predicted value was in good agreement with the experimental value. Akbardoost et al. [17] predicted the fracture resistance of samples with different geometric forms by MMTS criterion considering stress non-singular terms (T-stress, A_3, and B_3); the obtained results were in good agreement with the experimental values, and A_3 could be used to accurately calculate the length of the fracture process zone (FPZ). Yang et al. [18] studied the influence of water damage on the mechanical properties of rock fracture and found that the MMTS criterion was consistent with the test results when the damage degree was low, but it was inconsistent with the test results when the damage degree was deepened. Wang et al. [19] modified the GMTS criterion by considering parallel and vertical non-singular terms (T_x and T_y), and conducted numerical simulation based on the discontinuous deformation analysis (DDA) procedure. It was found that the simulation results were in good agreement with the test results, which verifies the accuracy of the modified GMTS criterion. Aliha et al. [20] studied the effect of loading rate on the composite (I+II) fracture behavior of short bend beam specimens and compared the prediction results of MMTS criterion with experimental results, showing that MMTS has a good accuracy in predicting fracture toughness. The radius of crack propagation plays an important role in predicting fracture behavior. Feng et al. [21,22] carried out different times of thermal cycle tests on granite at 20~300 °C and concluded that there was an error between the prediction results of MMTS and the results obtained by experimental back-calculation. Tang et al. [23–25] found in their study that the modified MTS criterion after introducing T-stress could capture the crack growth path more accurately; the growth path presented smooth characteristics, and the tensile fracture was enhanced with the increase in cohesion or the decrease in tensile strength.

The rock environment in underground engineering is very complex. Generally speaking, from the surface to the interior of the earth, the temperature rises by 20 °C to 30 °C for every 1 km of depth [26]. The development and operation process of geothermal reservoirs involves the complex spatio-temporal evolution law of temperature field, stress field, and hydrodynamic field in fractured rock mass [27–29]. Due to the different occurrence depths, the temperature and stress environment of geothermal reservoirs are different. In order to obtain more heat during geothermal extraction, geothermal reservoirs often require a sufficient fracture-network structure to provide more area for heat exchange. On the other hand, when reservoir fractures develop to a certain extent, uncontrolled fractures will cause damage to the surrounding strata, resulting in massive loss of injected fluids, engineering earthquakes, and other problems. Therefore, it is of great significance to deeply understand how different occurrence depths affect the fracture mechanical properties of geothermal reservoirs. Some scholars have studied the physical and mechanical properties of rocks at different occurrence depths. Zhang et al. [30] studied the AE sequence characteristics of coal in different damage and failure processes, and the research showed that the AE signal intensity of coal in the temperature pressure coupling process was higher than that under uniaxial compression and heating. Zhang et al. [31] found that the use of multi-source data and an integrated model will significantly improve the accuracy of wellbore instability and leakage prediction in high-pressure, high-temperature (HPHT) gas fields. Wang et al. [32] studied the properties of Jinping marble, Bayu granite, and Longchang sandstone under the coupling conditions of HPHT. The results show that the peak strength, elastic modulus, and fracture mode of rock vary with the change of rock type and temperature. Guo et al. [33] established a thermo-hydro-mechanical-damage (THMD) model to study the evolution of multiple physical fields during HDR hydraulic fracturing. Kumari et al. [34] carried out hydraulic fracturing tests on rocks under a confining pressure of 0~60 MPa and a temperature of ~300 °C. The study showed that the reservoir failure pressure increased linearly with the increase in confining pressure and decreased linearly with the increase in temperature. Meng et al. [35] conducted triaxial cyclic loading and unloading tests on rock specimens treated with different temperatures. The study showed that the increase in temperature could significantly degrade the physical properties of rock, and the increase in confining pressure could improve the strength and deformation of rock. Ding et al. [36] conducted a CO_2 seepage test on sandstone under the action of thermodynamic coupling, and the study showed that the permeability of sandstone decreased with the rise in stress, and first decreased and then increased with the increase in temperature.

It can be seen that, with its accuracy, MMTS can be used to predict the fracture characteristics of rock. However, how accurate is MMTS rock prediction considering the influence of different occurrence depths of geothermal reservoirs? Further research is needed. Meanwhile, there is a lack of research considering the failure and fracture mechanical behavior of reservoir rocks under the influence of geothermal reservoir depth. In fact, complex physical and chemical reactions occur in rock at different occurrence depths (temperature and three-dimensional stress environment), which affect the structure and properties of rock and inevitably affect the fracture process zone (FPZ) at rock fracture tips. Therefore, when MMTS is used to predict rock fracture characteristics, the influence of thermo-mechanical coupling on rock physical and mechanical properties should be considered.

Therefore, in this paper, the modified maximum tangential stress (MMTS) criterion is first analyzed in this study. The granite samples were then subjected to heat treatment under different temperatures and pressures, and the fracture toughness and tensile strength were tested. Finally, the prediction results of MMTS are compared with the experimental results, and the variation of the critical crack propagation radius (r_c) with temperature and three-dimensional stress is analyzed. The accuracy of MMTS in predicting the fracture characteristics of geothermal reservoir rocks considering the influence of different occurrence depths is revealed.

2. The Modified Maximum Tangential Stress (MMTS) Criterion

According to the MTS criterion proposed by Erdogan and Sih, when the shear stress, σ_θ, is equal to the tensile strength of the material, a crack is formed along the maximum tangential stress (θ_0) from the crack tip [37]. Mirsayar et al. [38] pointed out that considering the influence of higher-order terms higher than T-stress in practical engineering applications would increase the difficulty of solving the problem, so only the role of T-stress should be studied in higher-order terms. Therefore, according to the Williams infinite series expansion [39], not only the effect of the singular term of stress component, namely stress intensity factor, but also the effect of the non-singular constant term T-stress on crack propagation were considered, so the tangential stress component at the crack tip could be expressed as follows:

$$\sigma_\theta = \frac{1}{\sqrt{2\pi r}} \cos^2 \frac{\theta}{2} \left(K_I \cos \frac{\theta}{2} - 3K_{II} \sin \frac{\theta}{2} \right) + T \sin^2 \theta \tag{1}$$

where θ and r are the traditional crack tip coordinates, as shown in Figure 1, K_I and K_{II} are Mode I and Mode II stress intensity factors, respectively, and T is T-stress, which is independent of the distance to the crack tip [40].

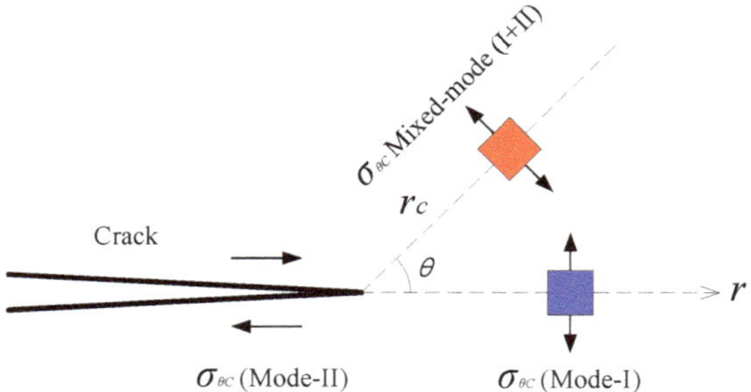

Figure 1. The maximum tangential stress (MTS) criterion [41,42].

The calculation formula of *T*-stress is as follows [43]:

$$T = \frac{P}{2RB} T * \left(\frac{a}{R}, \frac{S}{D}, \alpha \right) \tag{2}$$

where *T** is a dimensionless geometric parameter, which is related to fracture length ratio, span ratio, and fracture inclination angle [21,22]. By differentiating σ_θ with respect to θ, the crack initiation angle, θ_0, of mixed-mode (I+II) type can be obtained. The formula is as follows:

$$\frac{\partial \sigma_\theta}{\partial \theta} = 0, \; [K_I \sin\theta_0 + K_{II}(3\cos\theta_0 - 1)] - \frac{16T}{3}\sqrt{2\pi r_c}\cos\theta_0 \sin\frac{\theta_0}{2} = 0 \tag{3}$$

where r_c is the length of the fracture process zone (FPZ), which is regarded as an inherent characteristic of the material and has nothing to do with the shape of the sample, loading conditions and mixing mode [44]; r_c indicates that the fracture process zone has been damaged and cannot bear any load [38]. Schmidt proposed an equation based on the maximum normal stress theory to calculate r_c [45]:

$$r_c = \frac{1}{2\pi}\left(\frac{K_{IC}}{\sigma_t}\right)^2 \tag{4}$$

where K_{IC} is pure Mode I fracture toughness and σ_t is tensile strength, which can be calculated by Brazilian splitting experiment; r_c can be obtained from Equation (4) and θ_0 can be obtained by substituting it into Equation (3). Substituting r_c and θ_0 into Equation (1), the calculation formula of critical tangential stress, $\sigma_{\theta C}$, under mixed-mode (I+II) type can be obtained, which can be used to predict the fracture of mixed-mode (I+II) type brittle materials or quasi-brittle materials. The calculation formula is as follows:

$$\sigma_{\theta\theta}\sqrt{2\pi r_c} = \cos^2\frac{\theta_0}{2}\left(K_I\cos\frac{\theta_0}{2} - 3K_{II}\sin\frac{\theta_0}{2}\right) + \sqrt{2\pi r_c}\,T\sin^2\theta_0 \tag{5}$$

When pure Mode I fracture occurs, $\theta_0 = 0$, which satisfies:

$$K_{IC} = \sigma_{\theta\theta}\sqrt{2\pi r_c} \tag{6}$$

Based on Equations (5) and (6), the relationship between mixed-mode (I+II) fracture toughness, K_I and K_{II}, and pure Mode I fracture toughness, K_{IC}, can be established as follows:

$$K_{IC} = \cos^2\frac{\theta_0}{2}\left(K_I\cos\frac{\theta_0}{2} - 3K_{II}\sin\frac{\theta_0}{2}\right) + \sqrt{2\pi r_c}\,T\sin^2\theta_0 \tag{7}$$

The effective stress intensity factor can be calculated as follows:

$$K_{eff} = \sqrt{K_I^2 + K_{II}^2} \tag{8}$$

The effective fracture toughness and pure Mode I fracture toughness ratio, K_{eff}/K_{IC}, can be obtained from Equations (7) and (8), and its theoretical expression is as follows [18]:

$$\frac{K_{eff}}{K_{IC}} = \sqrt{Y_I^2 + Y_{II}^2}\left[\cos^2\frac{\theta_0}{2}\left(Y_I\cos\frac{\theta_0}{2} - 3Y_{II}\sin\frac{\theta_0}{2}\right) + T * \sqrt{\frac{2\pi r_c}{a}}\sin^2\theta_0\right]^{-1} \quad (9)$$

where Y_I is the normalized Mode I stress intensity factor and Y_{II} is the normalized Mode II stress intensity factor. The fracture toughness can be expressed by the mixed-mode parameter, M^e, whose theoretical expression is as follows [21,22]:

$$M^e = \frac{2}{\pi}\tan^{-1}\frac{K_I}{K_{II}} \quad (10)$$

where the value of M^e ranges from 0 to 1. When the pure Mode I fracture occurs, the value of M^e is 1. When the pure Mode II fracture occurs, the M^e value is 0. While mixed-mode (I+II) fracture occurs, the value of M^e is related to the fracture dip angle.

Under the action of external load, stress concentration occurs at the fracture tip. The initiation, propagation, interaction, and aggregation of internal microcracks are crucial reasons for the instability and failure of rock. The stress condition is affected by irregular crystal structure, original pores and cracks, friction force, and other factors [46], which will lead to high shear and tensile stress at the fracture tip. When stress exceeds the strength of molecular and atomic bonds, chemical bonds break and microcracks form [47]. Before reaching the peak load, microcracks constantly sprout at the crack tip, and these microcracks nucleate and propagate to form the microcrack expansion area, namely the fracture process zone (FPZ) of rock [18]. When the peak load is reached, the crack tip begins to crack and expands, finally forming macroscopic cracks.

FPZ is an important fracture mechanical parameter of rocks. Its size is closely related to the size of specimens and the properties of rocks, such as fracture toughness, fracture trajectory, and fracture energy [47]. r_c can be regarded as the length of FPZ. When r_c is changed, it can be seen from Equation (3) that the fracture initiation angle, θ_0, of rock is changed, and from Equation (9) that the predicted value of mixed-mode (I+II) fracture toughness is changed. In conclusion, FPZ seriously affects fracture initiation angle and mixed-mode (I+II) fracture toughness, which further affects the accuracy of MMTS in predicting rock fracture behavior. If the value of crack growth radius, r_c, can be known, the characteristics such as crack growth angle, fracture toughness value, and fracture trajectory can be predicted by MMTS criterion.

3. Methodology

In order to draw the fracture envelope of MMTS, it can be seen from Equations (4), (9), and (10) that the values of tensile strength and the Mode I, Mode II, and mixed-mode (I+II) fracture toughness of rock should be tested. Therefore, in this section, we first introduce the rock samples used in this experiment. We will then introduce the thermal and mechanical experiments, and finally, we will introduce the mechanical experiments.

The granite samples selected in this paper were taken from Miluo, Hunan Province. Through X-ray diffraction and thin section identification, the granite samples used in this experiment are mainly composed of quartz, plagioclase, alkaline feldspar, biotite, and Muscovite [48].

The geothermal reservoir is located in different temperature and three-dimensional stress in situ stress environments at various occurrence depths. The samples are set under different temperature and three-dimensional pressure conditions for heat treatment. Six groups of experimental conditions were set: the first group was set as axial pressure of 25 MPa, lateral pressure of 20 MPa, and temperature of 100 °C. In the second group, the axial pressure was 40 MPa, the lateral pressure was 30 MPa, the temperature was 130 °C. In the third group, the axial pressure was 50 MPa, the lateral pressure was 35 MPa, and temperature was 100 °C. In the fourth set, the axial pressure was 50 MPa, the lateral pressure was 40 MPa, and the temperature was 150 °C. In the fifth set, the axial pressure was 60 MPa, the lateral pressure was 40 MPa, and the temperature was 100 °C. In the sixth set, the axial pressure was 75 MPa, the lateral pressure was 50 MPa, and the temperature was 200 °C. Heat treatment experiments of granite were carried out using the multifunctional high temperature rock triaxial servo control permeation testing machine of Taiyuan University of Technology. During the test, the cylinder sample was first placed into the heating furnace, and the pressure was applied to the set value at the axial loading rate of 0.5 kN/s and the lateral loading rate of 0.02 mm/min, and it was then heated at the heating rate of 5 °C/min. When the set temperature was reached, the heat was held for two hours, and then the heating was stopped and the pressure was reduced. The sample was allowed to cool naturally to room temperature [48].

The granite samples after the heat treatment tests were processed by wire cutting into standard Brazilian disc and standard semi-circular bend (SCB) samples, which were used for tensile strength tests and fracture toughness tests, respectively. Using the International Society for Rock Mechanics (ISRM)-recommended standard SCB sample geometry size range [49], the sample processing parameters can be determined. In this experiment, the diameter (D) of the sample is 50 mm, the thickness (B) is 20 mm, the length of the artificial groove (a) is 12.5 mm, and the span of the support (S) is 30.5 mm. Various loading modes can be obtained by changing the angle (α) between the pre-cut notch and the loading direction. When $\alpha = 0°$, it belongs to Mode I fracture. When $\alpha = 54°$, it belongs to Mode II fracture. In this study, $\alpha = 30°$ were selected as the mixed-mode (I+II) fracture mode.

The three point bending fracture mechanics test was then carried out on the SCB samples. An INSTRON 5967 testing machine of Sichuan University was used to test the fracture toughness of the SCB samples under mechanical loading. During the test, the displacement-controlled loading mode was adopted, and the loading rate was 0.002 mm/s. At the same time, the rock mechanics test system MTS815 Flex Test GT model of Sichuan University was used to test the tensile strength of the Brazilian disc samples after heat treatment. The loading rate was 0.1 mm/min. The experimental process is shown in Figure 2.

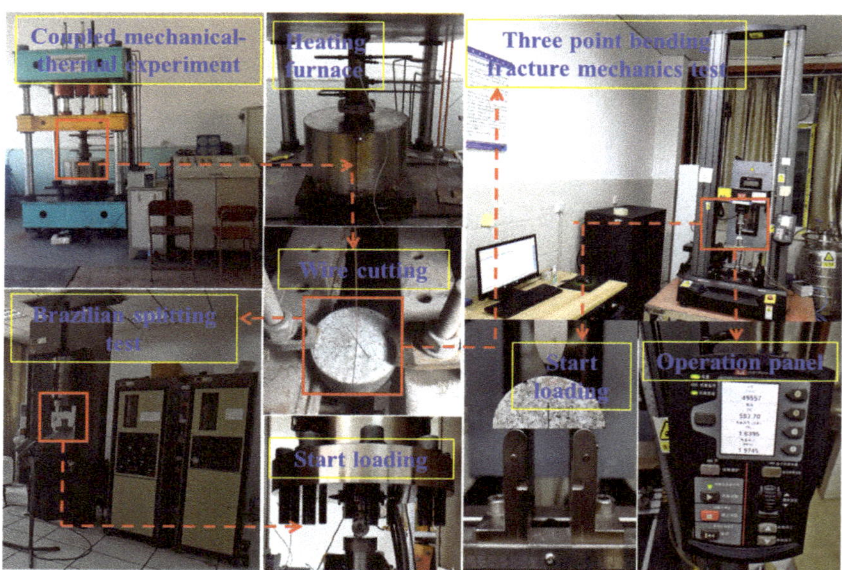

Figure 2. Test flow chart (Some pictures were modified from reference [48]).

4. Experimental Results

In Section 3, the experimental results of the fracture toughness and tensile strength of granite specimens affected by different thermal and force treatment conditions are obtained after the thermal-mechanical treatment test, three-point bending fracture mechanics test, and Brazil splitting test.

4.1. Test Results of Fracture Toughness

In Section 3, the thermal-mechanical treatment test and three-point bending fracture mechanics test were carried out on granite specimens. The average value of the obtained experimental results is shown in Table 1.

Table 1. Average values of the effective fracture toughness of granite.

Group Number	Conditions of Experiment	Fracture Mode	K_{eff} (MPa·m$^{1/2}$)
Group 1	The axial pressure is 25 MPa, the lateral pressure is 20 MPa, and the temperature is 100 °C	Mode I Mixed mode (I+II) ($\alpha = 30°$) Mode II	1.396 1.086 0.437
Group 2	The axial pressure is 40 MPa, the lateral pressure is 30 MPa, and the temperature is 130 °C	Mode I Mixed mode (I+II) ($\alpha = 30°$) Mode II	1.344 1.137 0.532
Group 3	The axial pressure is 50 MPa, the lateral pressure is 35 MPa, and the temperature is 100 °C	Mode I Mixed mode (I+II) ($\alpha = 30°$) Mode II	1.536 0.915 0.416
Group 4	The axial pressure is 50 MPa, the lateral pressure is 40 MPa, and the temperature is 150 °C	Mode I Mixed mode (I+II) ($\alpha = 30°$) Mode II	1.481 0.949 0.407
Group 5	The axial pressure is 60 MPa, the lateral pressure is 40 MPa, and the temperature is 100 °C	Mode I Mixed mode (I+II) ($\alpha = 30°$) Mode II	1.581 0.969 0.381
Group 6	The axial pressure is 75 MPa, the lateral pressure is 50 MPa, and the temperature is 200 °C	Mode I Mixed mode (I+II) ($\alpha = 30°$) Mode II	1.357 0.925 0.461

4.2. Brazilian Disk Splitting Experiment Results

In Section 3, the thermo-mechanical treatment test and the Brazilian disk splitting mechanics test were carried out on granite specimens. After data processing, the tensile strength results of granite under different thermo-mechanical treatment conditions were obtained. Concurrently, the Brazilian disk splitting test is often a conventional experimental method to study the mechanical properties of rocks. Therefore, the tensile strength and the Brazilian splitting modulus obtained from the Brazilian splitting experiment were studied. These were used to explore the influence of different occurrence depths on the mechanical properties of granite reservoirs.

4.2.1. Tensile Strength

The tensile strength of granite under different temperature and three-dimensional stress is shown in Figure 3.

According to Figure 3a, the Brazilian splitting strength, σ_t, of granite fluctuates with the increase in temperature. The Brazilian splitting strength at 150 °C reaches the maximum value (9.451 MPa), which is 28% higher than that of the untreated sample. According to Figure 3b, with the increase in deviatoric stress, the Brazilian splitting strength first increases, then decreases, and then increases. When the deviatoric stress is 5 MPa, the Brazilian splitting strength reaches the maximum value (9.122 MPa) and then decreases gradually with the increase in the deviatoric stress. When the deviatoric stress increases to 20 MPa, the Brazilian splitting strength of granite decreases to the minimum value (5.824 MPa), which is 21% lower than that of the untreated sample. When the deviatoric stress is 10 MPa, the Brazilian splitting strength of granite is 7.873 MPa at 130 °C and 9.451 MPa at 150 °C, increasing by 20%. When the temperature is 100 °C, the Brazilian splitting strength of granite is 9.122 MPa when the deviatoric stress is 5 MPa and 5.824 MPa when the deviatoric stress is 20 MPa, decreasing by 36%. Therefore, with the change of external temperature and pressure, the Brazilian splitting strength of granite will not continue to decrease, but will increase.

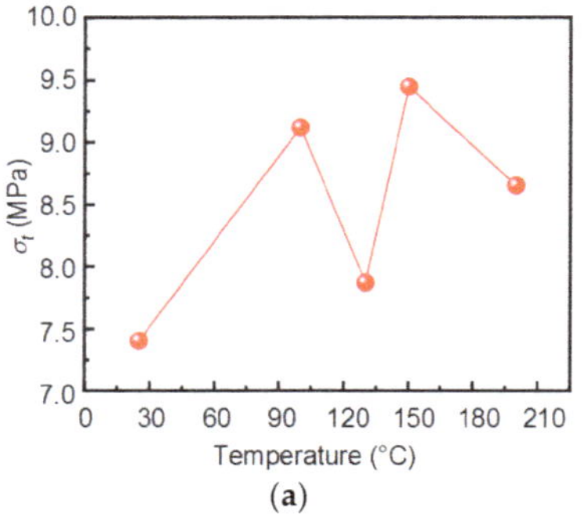

(**a**)

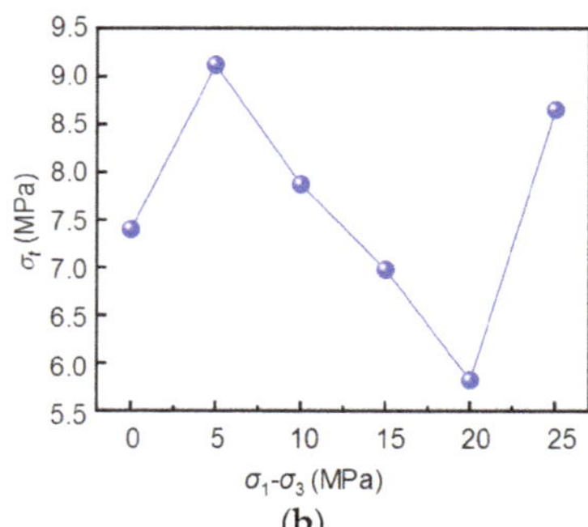

(**b**)

Figure 3. Brazilian splitting strength curves under different heat treatment conditions. (**a**) Brazilian splitting strength curve with temperature; (**b**) Brazilian splitting strength curve with deviatoric stress.

4.2.2. Brazilian Splitting Modulus

The ability of rock to resist elastic deformation and failure can be determined by the Brazilian splitting experiment. Brazilian splitting modulus (E_T) was used as the benchmark for judgment. The specific method is as follows: divide the ordinate of the load-displacement curve by the area of the meridional plane of the rock and the abscate by the diameter of the specimen to obtain the slope of the elastic phase, namely the Brazilian splitting modulus (E_T) [50]. The Brazilian splitting modulus curve of the sample was thus drawn, as shown in Figure 4:

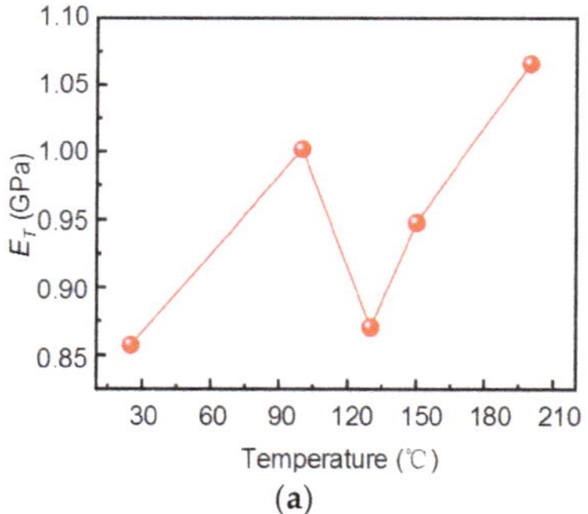

(**a**)

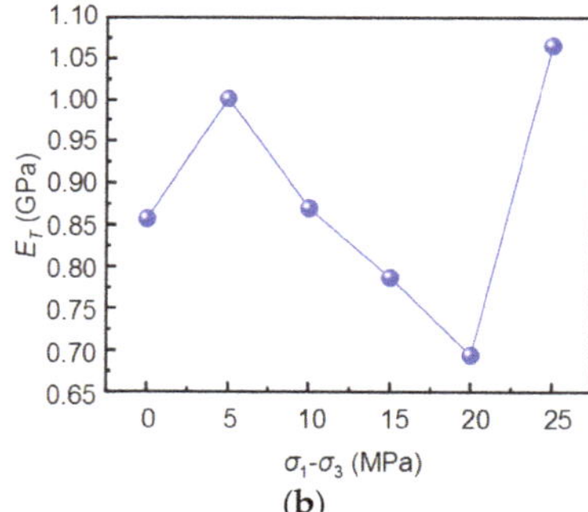

(**b**)

Figure 4. Variation curve of Brazilian splitting modulus of granite. (**a**) Brazilian splitting modulus variation curve with temperature; (**b**) Brazilian splitting modulus variation curve with deviatoric stress.

As can be seen from Figure 4a, with the temperature increasing, the Brazilian splitting modulus (E_T) first increases, then decreases, and then increases, which is generally higher than the Brazilian splitting modulus at room temperature. When the Brazilian splitting modulus (E_T) reaches its maximum value (1.07 GPa) at 200 °C, which is 24% higher than that at room temperature, the rock has the strongest ability to resist elastic deformation and failure. Pressure compacts the pores and cracks inside the rock, offsets the deterioration of rock properties caused by temperature, and improves the mechanical properties of the rock. As can be seen from Figure 4b, with the increase in deviatoric stress, the Brazilian splitting modulus generally shows an upward-down-upward trend, which is consistent with the variation trend of Brazilian splitting strength with the escalation of deviatoric stress. When the deviatoric stress is 20 MPa, the Brazilian splitting modulus reaches the minimum value (0.69 GPa), which decreases by 19% compared with the room temperature condition, indicating that the thermal damage inside the rock is the most serious and the resistance to elastic deformation and failure is the weakest under this condition. The Brazilian splitting modulus decreases by 12% when deviatoric stress increases from 15 MPa to 20 MPa when the temperature is 100 °C. Accordingly, with the change of external temperature and pressure, the Brazilian splitting modulus of granite will increase, and appropriate temperature and pressure can improve the ability of rock to resist elastic deformation.

5. Modification of Granite MMTS Criterion Considering Different Occurrence Depths

In Section 4, the fracture toughness and tensile strength of granite affected by different occurrence depths are obtained. We substituted these parameters into the theoretical formula in Section 2 to study the accuracy of MMTS in predicting rock fracture characteristics.

5.1. Comparative Analysis of Experimental Results and MMTS Criterion Prediction Results

When Equation (7) is combined with the calculation formulas of K_I and K_{II}, the following formula can be obtained:

$$\frac{K_{IC}}{K_I} = \cos^2 \frac{\theta_0}{2} \left(\cos\theta_0 - 3\frac{Y_I}{Y_{II}} \sin \frac{\theta_0}{2} \right) + \frac{1}{Y_I} \sqrt{\frac{2\pi r_c}{a}} T * \sin^2 \theta_0 \tag{11}$$

$$\frac{K_{IC}}{K_{II}} = \cos^2 \frac{\theta_0}{2} \left(\frac{Y_I}{Y_{II}} \cos\theta_0 - 3 \sin \frac{\theta_0}{2} \right) + \frac{1}{Y_{II}} \sqrt{\frac{2\pi r_c}{a}} T * \sin^2 \theta_0 \tag{12}$$

According to Equations (11) and (12), a relationship between K_I/K_{IC} and K_{II}/K_{IC} can be established, and the fracture envelope under different temperatures and pressures can be obtained, as shown in Figure 5. As can be seen from Figure 5, with the rise in temperature and deviatoric stress, the fracture envelope shows obvious change characteristics, which indicates that various temperatures and pressures have different effects on the fracture mechanical properties of rocks.

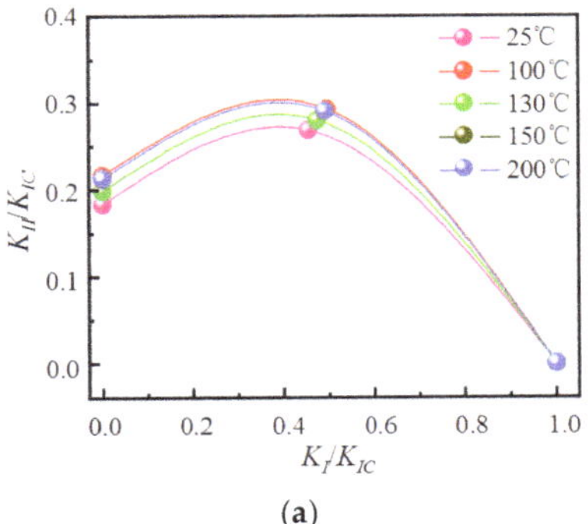

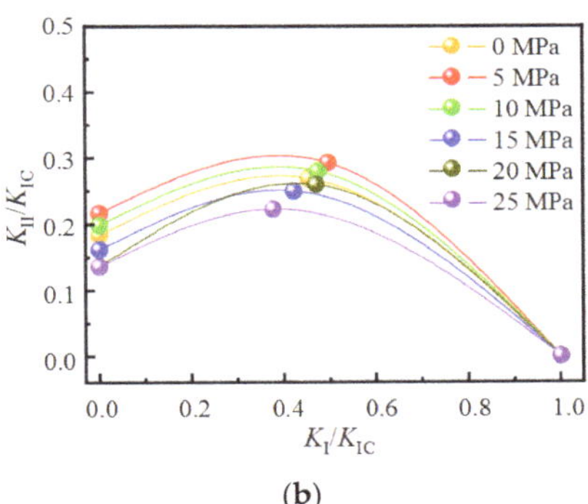

Figure 5. Fracture envelope under different conditions. (**a**) Fracture envelope at different temperatures; (**b**) Fracture envelope under different deviatoric stresses.

By combining Equations (9) and (10), the relationship between the mixed-mode parameter M^e and K_{eff}/K_{IC} is established. Figure 6 shows the comparison between the fitted curves of K_{eff}/K_{IC} and the predicted values of the traditional MTS theory and MMTS theory under various temperatures and pressures.

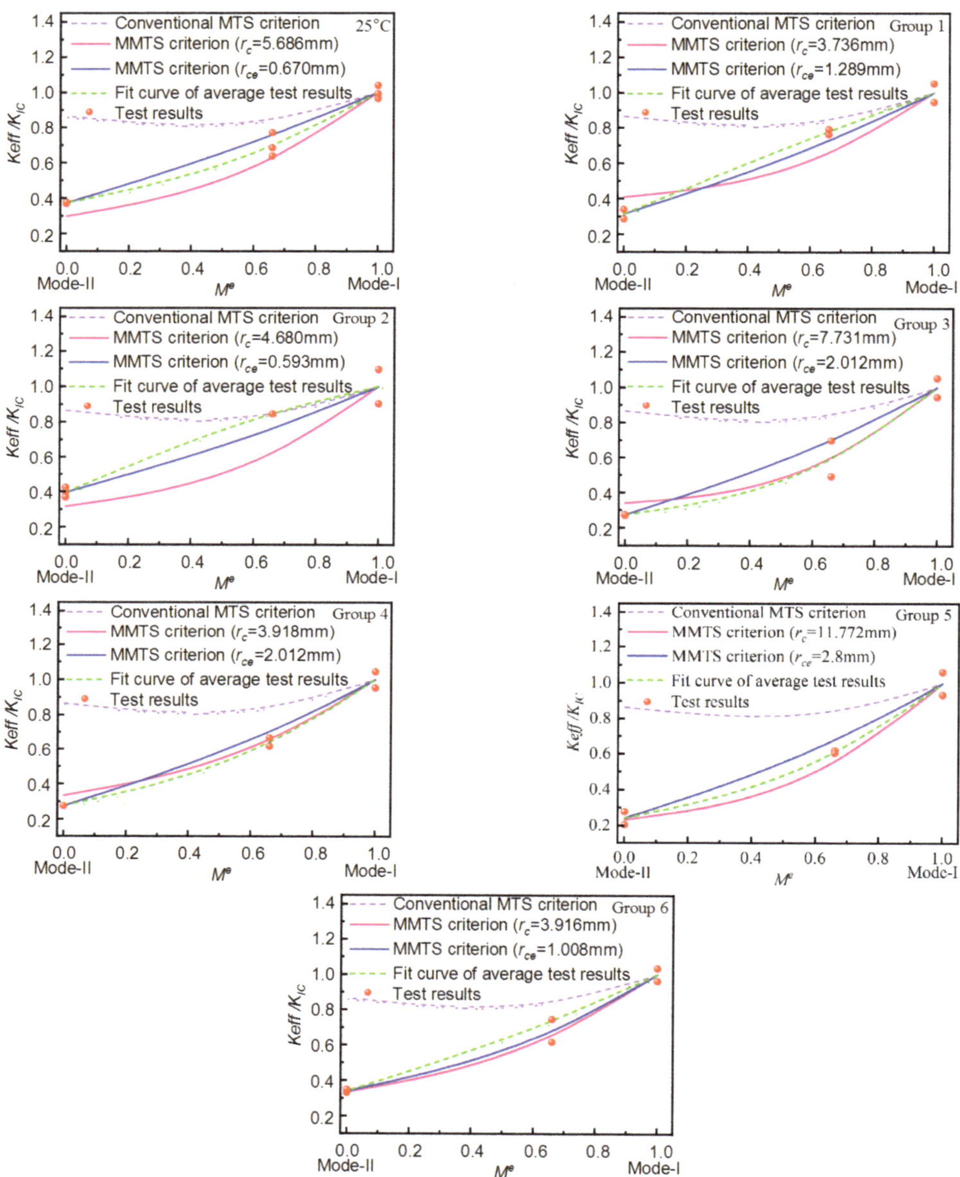

Figure 6. Comparison of the K_{eff}/K_{IC} ratio of granite SCB specimens with the theoretical value predicted by MMTS criterion.

It can be seen from Figure 6 that the theoretical prediction curve of the traditional MTS criterion deviates greatly from the test results, which may be mainly because the influence of T-stress is ignored. However, under different temperatures and pressures, the prediction curve of MMTS criterion after considering T-stress deviates greatly from the fitting curve of the average value of test results, and there is a large deviation from the experimental results. This indicates that the prediction of the Mode II fracture toughness and mixed-mode (I+II) fracture toughness of granites with different occurrence depths by MMTS has bigger differences than the actual results. Next, the critical crack propagation radius will be deduced from the experimental results to better match the actual results of Mode II and mixed-mode (I+II) fracture toughness of granite. In addition, it is worth noting that the fracture envelope at different occurrence depths is distinguished, which indicates that the influence of various occurrence depths should be considered when predicting rock fracture toughness by fracture criteria.

5.2. The Modified MMTS Criteria

It can be seen from the analysis in Section 2 that the crack propagation radius, r_c, and fracture initiation angle, θ_0, seriously affect the accuracy of MMTS criterion prediction. The modified critical crack propagation radius was defined as r_{ce}. By substituting the experimental results, K_{eff}/K_{IC}, into Equation (9), Equations (3) and (9) are simultaneously established to obtain the critical crack propagation radius, r_{ce}, which is in line with the actual experimental situation. By substituting r_{ce} into Equations (3) and (9), the modified MMTS prediction, K_{eff}/K_{IC}, curve is drawn, as shown by the solid blue line in Figure 6. It can be seen from Figure 6 that the MMTS prediction curve obtained by direct reverse calculation through experimental testing is in high agreement with the test results. This indicates that, compared with the r_c calculated by Equation (4), r_{ce} can better reflect the size of FPZ of granite samples with different occurrence depths. It can also be seen that the previous method of estimating the critical crack propagation radius directly by Equation (4) is not consistent with the actual situation. Maybe the fracture criterion needs to be further refined. Under the current MMTS theory, a new method for estimating or testing the critical crack propagation radius should be explored, or the r_c estimated by the previous method should be modified. r_{ce}/r_c is defined as the correction coefficient of crack propagation radius, which is expressed as γ. The changes of r_c, r_{ce}, and $\gamma(r_{ce}/r_c)$ with temperature and deviatoric stress were plotted, as shown in Figure 7.

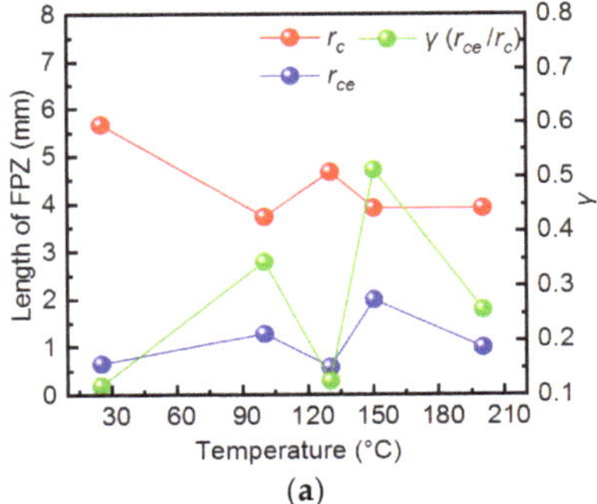
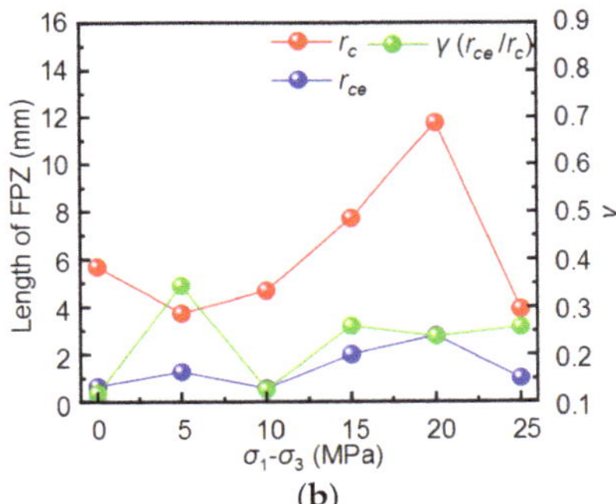

Figure 7. Changes of r_c, r_{ce}, and $\gamma(r_{ce}/r_c)$. (**a**) Variation of r_c, r_{ce}, and $\gamma(r_{ce}/r_c)$ with temperature. (**b**) Variation of r_c, r_{ce}, and $\gamma(r_{ce}/r_e)$ with deviatoric stress.

As can be seen from Figure 7, there are great differences in r_c and r_{ce} values; under different heat treatment conditions, r_{ce} is smaller than r_c. With the increase in temperature and deviational stress, the changes of r_c and r_{ce} are not coordinated, and $\gamma(r_{ce}/r_c)$ shows an upward trend, ranging from 0.118 to 0.514. This indicates that the critical crack propagation radius, r_c, obtained by Equation (4) cannot accurately reflect the size of FPZ. According to Section 4.1, the test results of fracture toughness under different heat treatment conditions and different modes have obvious differences. Therefore, r_c seriously affects the accuracy of MMTS criterion in predicting rock fracture toughness.

6. Numerical Simulation of Fracture of Granite Specimen

It can be seen from Section 5 that FPZ size affects the accuracy of MMTS criterion in predicting the fracture mechanical behavior of granite with different occurrence depths. In order to verify that FPZ is a critical reason for the accuracy of MMTS criterion in predicting granite fracture trajectory, we used ABAQUS numerical software and, based on the extended finite element method (XFEM), conducted numerical simulations for specimens with different occurrence depths and different crack inclination angles. The model was set as a two-dimensional plane stress model, and the maximum principal stress and damage variable were selected as the criterion for crack initiation and propagation [51]. Due to space limitation, Figure 8 shows the numerical simulation results with typical r_{ce} data.

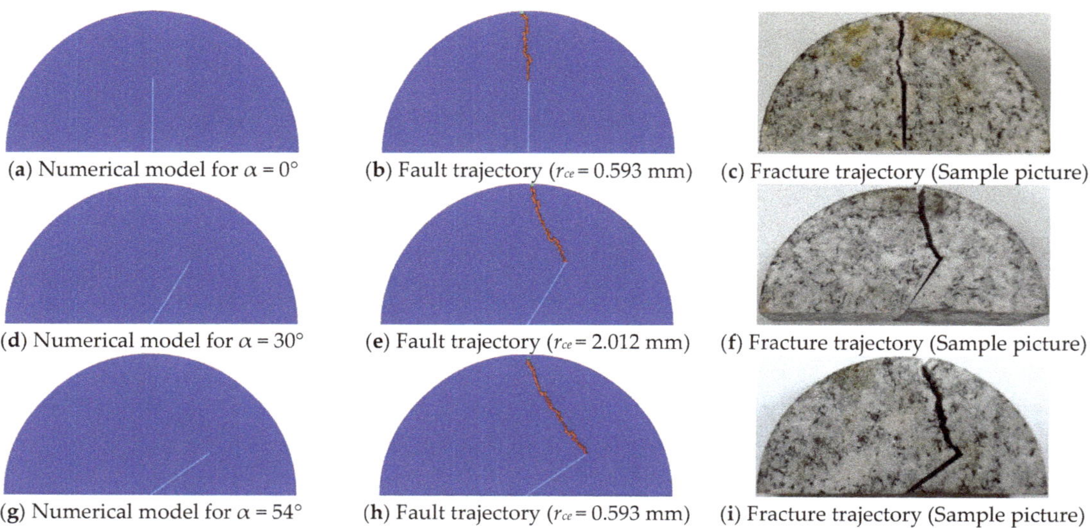

(**a**) Numerical model for $\alpha = 0°$ (**b**) Fault trajectory ($r_{ce} = 0.593$ mm)   (**c**) Fracture trajectory (Sample picture)

(**d**) Numerical model for $\alpha = 30°$ (**e**) Fault trajectory ($r_{ce} = 2.012$ mm)   (**f**) Fracture trajectory (Sample picture)

(**g**) Numerical model for $\alpha = 54°$ (**h**) Fault trajectory ($r_{ce} = 0.593$ mm)   (**i**) Fracture trajectory (Sample picture)

Figure 8. Comparison between the results of granite fracture trajectories after correction of critical crack propagation radius and the experimental results.

It can be seen from Figure 8 that the crack expands along the crack tip and the loading point in the simulation results, and the crack initiation angle increases with the increase in the prefabricated crack angle, which is consistent with the test results. The results of granite crack growth trajectory obtainedusing the modified critical crack growth radius, r_{ce}, are in good agreement with the experimental results. The numerical simulation results show that temperature and three-dimensional pressure affect the fracture trajectory of granite by influencing FPZ, which is an important factor affecting the prediction accuracy of MMTS criterion. Therefore, it is crucial to accurately find the critical crack propagation radius for accurately predicting the Mode II fracture toughness, mixed-mode (I+II) fracture toughness, and fracture trajectory of granite with different occurrence depths.

7. Discussion

The temperature and geostress of geothermal reservoirs with different occurrence depths vary, and the mechanical properties of rock are different. Temperature and pressure do not affect rocks in the same way. As the rock is composed of multiple mineral components and the mineral particles have different thermal expansion coefficients, the thermal expansion between mineral particles is not coordinated in the process of heating and insulation, resulting in thermal stress among mineral particles. When the thermal stress is large enough, it can promote the expansion of the primary cracks and produce microcracks [52–54], weakening the mechanical properties of the rock. In addition, the water inside the rock evaporates during the heating process, producing more micropores, reducing the compactness of the rock, thereby weakening the mechanical properties of the rock. On the contrary, the appropriate pressure can facilitate pore closure, suppress the propagation of primary fractures and new fractures of the rock, weaken the thermal expansion of the rock, make the rock denser, and further improve the mechanical properties of the rock [55,56]. Therefore, due to the different dominant positions of temperature and pressure, the mechanical properties of rocks will increase or decrease.

The experimental results show that under different temperature pressure coupling mechanisms, the fracture toughness, Brazilian splitting strength, and Brazilian splitting modulus of rock fluctuate up and down with the increase in confining pressure and temperature, showing obvious nonlinear characteristics. For example, when the temperature is 100 °C, the axial pressure is 25 MPa, and the lateral pressure is 20 MPa, the mixed-mode (I+II) fracture toughness, Brazilian splitting strength, and the Brazilian splitting modulus of the rock are higher than those at room temperature. Meng et al. [57] conducted triaxial compression tests on limestone under high temperature. The results showed that the peak stress of rock increased with the increase in confining pressure from 0 MPa to 30 MPa at a certain temperature. Funatsu et al. [58] tested the fracture toughness of Kimachi sandstone under the coupling of temperature and pressure. The results show that the effective fracture toughness increases with the increase in confining pressure from 1 MPa to 5 MPa when the temperature is constant. The above research results show that appropriate pressure can counteract the deterioration of rock mechanical properties caused by temperature and improve the mechanical properties of rock. When the temperature is 100 °C, the axial pressure is 60 MPa, and the lateral pressure is 40 MPa, the mixed-mode (I+II) fracture toughness, Brazilian splitting strength, and Brazilian splitting modulus of the rock are lower than those at room temperature. This may be due to the severe crystal deformation and slip in the rock caused by excessive three-dimensional pressure, and the internal fracture is more serious, leading to the reduction of the mechanical properties of the rock. Based on the above analysis, it can be seen that under different occurrence depths, the ability of rock to resist tension, shear, and tension shear fracture is different.

The size of the fracture process zone (FPZ) is closely related to the internal microstructure of the rock. FPZ is not only formed in homogeneous and dense media, but also under temperature and pressure damage, its original state should contain some pores and cracks. It is known from the test results that under the action of temperature and three-dimensional pressure, the crack growth radius before and after correction fluctuates and decreases with the increase in temperature. Meng et al. [47] carried out fracture mechanics experiments under different temperatures and found that FPZ value gradually increases with the increase in temperature. There are significant differences compared with this research. This is mainly because, under the combined action of temperature and pressure, complex physical and chemical reactions occur in the rock. In addition to the role of thermal stress, three-dimensional pressure can promote the closure of pores and fractures in the rock and inhibit the initiation and propagation of microcracks, and the FPZ value of granite will also be affected.

The modified MMTS criterion can accurately predict the fracture characteristics of geothermal reservoirs with different occurrence depths. According to Equations (3)–(9), the crack growth radius is closely related to the crack growth angle, mixed-mode (I+II), and Mode-II fracture toughness parameters, and affects the macro fracture characteristics. It can be seen from Section 5 that the modified crack growth radius (r_{ce}) can accurately predict the fracture toughness of rock compared with r_c. This shows that r_{ce} can better reflect the FPZ size of rock. In addition, some scholars [21,22,47,59] also pointed out that the actual FPZ size calculated by Equation (4) was inconsistent. Therefore, it is necessary to correct r_c when predicting the fracture characteristics of geothermal reservoirs with different depths.

8. Conclusions

In this paper, the accuracy of MMTS in predicting the fracture mechanical behavior of geothermal reservoirs is analyzed. Firstly, the theoretical formula is derived, and then the MMTS criterion is analyzed and modified by heat treatment, three-point bending fracture, and Brazilian splitting tests of granite samples in a geothermal reservoir under different temperatures and three-dimensional pressures. The main research conclusions are as follows:

(1) FPZ seriously affects fracture initiation angle and mixed-mode (I+II) fracture toughness, which affects the accuracy of MMTS in predicting rock fracture behavior. Considering the influence of different occurrence depths, r_c fails to accurately reflect the actual size of FPZ, so it needs to be modified to obtain results consistent with the actual. The Mode II fracture toughness and mixed-mode (I+II) fracture toughness of granites with different occurrence depths predicted by MMTS are significantly different from the measured values.

(2) The r_{ce} obtained from the reverse solution of the test results is smaller than the r_c. With the increase in temperature and deviatoric stress, the changes of r_c and r_{ce} are not coordinated. When the r_{ce} obtained by the reverse calculation of the experimental data is substituted into the MMTS criterion, the predicted results are in good agreement with the experimental results. Temperature and three-dimensional stress affect the size of FPZ by changing the microscopic properties of rock. It is also verified by numerical simulation that after correcting the critical crack propagation radius, the fracture trajectories obtained are basically consistent with the experimental results. Further theoretical and experimental studies are needed to evaluate the fracture characteristics of geothermal reservoirs at different depths predicted by MMTS criteria.

(3) Under the influence of temperature and pressure, the Brazilian splitting strength, σ_t, of granite fluctuates and rises with the increase in temperature. With the increase in deviatoric stress, Brazilian splitting strength and Brazilian splitting modulus first increase, then decrease, and then increase.

Author Contributions: Conceptualization, C.L. (Chenbo Liu) and G.F.; data curation, Z.D. and G.L.; formal analysis, J.W. and Y.T.; funding acquisition, G.F.; methodology, C.L. (Chenbo Liu) and G.F.; resources, H.X. (Hongqiang Xie), H.X. (Huining Xu), Y.H. (Yaoqing Hu), C.L. (Chun Li) and Y.H.; writing—review and editing, C.L. (Chenbo Liu), G.F., G.L., Y.H. (Yaoqing Hu), Y.H. (Yuefei Hu), Q.W. and L.C. All authors have read and agreed to the published version of the manuscript.

Funding: This study was funded by the National Natural Science Foundation of China (No. 52104143), the open fund of Key Laboratory of Deep Earth Science and Engineering (Sichuan University), Ministry of Education (Grant No. DESE202104), the Natural Science Foundation of Sichuan Province, China (Grant No. 52104143), the Fundamental Research Funds for the Central Universities (No. 2021SCU12039), the College Students' Innovative Entrepreneurial Training Plan Program (No. 20220708L), and China Postdoctoral Science Foundation (Grant No. 2020M673225). These supports are gratefully acknowledged.

Institutional Review Board Statement: Not applicable.

Informed Consent Statement: Not applicable.

Data Availability Statement: The data that support the findings of this study are available from the corresponding author upon reasonable request.

Conflicts of Interest: No conflict of interest exists in the submission of this manuscript.

References

1. Xie, H.; Li, C.; Zhao, T.; Chen, J.; Liao, J.; Ma, J.; Li, B. Conceptualization and evaluation of the exploration and utilization of low/medium-temperature geothermal energy: A case study of the Guangdong-Hong Kong-Macao Greater Bay Area. *Geomech. Geophys. Geo-Energy Geo-Resour.* **2020**, *6*, 18. [CrossRef]
2. Sun, Z.; Zhang, X.; Xu, Y.; Yao, J.; Wang, H.; Lv, S.; Sun, Z.; Huang, Y.; Cai, M.; Huang, X. Numerical simulation of the heat extraction in EGS with thermal hydraulic-mechanical coupling method based on discrete fractures. *Energy* **2017**, *120*, 20–33. [CrossRef]
3. Ye, Z.; Wang, J.; Hu, B. Comparative study on heat extraction performance of geothermal reservoirs with presupposed shapes and permeability heterogeneity in the stimulated reservoir volume. *J. Pet. Sci. Eng.* **2021**, *206*, 109023. [CrossRef]
4. Liu, B.; Suzuki, A.; Watanabe, N.; Ishibashi, T.; Sakaguchi, K.; Ito, T. Fracturing of granite rock with supercritical water for superhot geothermal resources. *Renew. Energy* **2022**, *184*, 56–67. [CrossRef]
5. Wang, G.; Zhang, W.; Ma, F.; Lin, W.; Liang, J.; Zhu, X. Overview on hydrothermal and hot dry rock researches in China. *China Geol.* **2018**, *1*, 273–285. [CrossRef]
6. Zhu, J.; Hu, K.; Lu, X.; Huang, X.; Liu, K.; Wu, X. A review of geothermal energy resources, development, and applications in China: Current status and prospects. *Energy* **2015**, *93*, 466–483. [CrossRef]
7. Ganguly, S.; Kumar, M.S.M. Geothermal reservoirs—A brief review. *J. Geol. Soc. India* **2012**, *79*, 589–602. [CrossRef]
8. Zhou, Z.; Jin, Y.; Zeng, Y.; Zhang, X.; Zhou, J.; Zhuang, L.; Xin, S. Investigation on fracture creation in hot dry rock geothermal formations of China during hydraulic fracturing. *Renew. Energy* **2020**, *153*, 301–303. [CrossRef]
9. Lu, S.M. A global review of enhanced geothermal system (EGS). Renew. *Sustain. Energy Rev.* **2018**, *81*, 2902–2921. [CrossRef]
10. Zhang, Y.; Zhao, G. A global review of deep geothermal energy exploration: From a view of rock mechanics and engineering. *Geomech. Geophys. Geo-Energy Geo-Resour.* **2019**, *6*, 4. [CrossRef]
11. Kang, F.; Tang, C.; Li, Y.; Li, T.; Men, J. Challenges and opportunities of enhanced geothermal systems: A review. *Chin. J. Eng.* **2022**, *44*, 1767–1777. (In Chinese)
12. Olasolo, P.; Juarez, M.C.; Morales, M.P.; D'Amico, S.; Liarte, I.A. Enhanced geothermal systems (EGS): A review. *Renew. Sustain. Energy Rev.* **2016**, *56*, 133–144. [CrossRef]
13. Richards, H.G.; Parker, R.H.; Green, A.S.P.; Jones, R.H.; Nicholls, J.D.M.; Nicol, D.A.C.; Randall, M.M.; Richards, S.; Stewart, R.C.; Willisrichards, J. The performance and characteristics of the experimental hot dry rock geothermal reservoir at rosemanowes, cornwall (1985–1988). *Geothermics* **1994**, *23*, 73–109. [CrossRef]
14. Seibt, P.; Hoth, P. The neustadt-glewe geothermal station: Form surveys to active operation. *Therm. Eng.* **2004**, *51*, 494–497.
15. Schill, E.; Genter, A.; Cuenot, N.; Kohl, T. Hydraulic performance history at the Soultz EGS reservoirs from stimulation and long-term circulation tests. *Geothermics* **2017**, *70*, 110–124. [CrossRef]
16. Ayatollahi, M.R.; Sistaninia, M. Mode-II fracture study of rocks using Brazilian disk specimens. *Int. J. Rock Mech. Min. Sci.* **2011**, *48*, 819–826. [CrossRef]
17. Akbardoost, J.; Ayatollahi, M.R. Experimental analysis of mixed mode crack propagation in brittle rocks: The effect of non-singular terms. *Eng. Fract. Mech.* **2014**, *129*, 77–89. [CrossRef]
18. Yang, J.; Liang, W.; Chen, Y.; Li, L.; Lian, H. Experiment research on the fracturing characteristics of mudstone with different degrees of water damage. *Chin. J. Rock. Mech. Eng.* **2017**, *36*, 2431–2440. (In Chinese)
19. Wang, C.; Wang, S. Modified generalized maximum tangential stress criterion for simulation of crack propagation and its application in discontinuous deformation analysis. *Eng. Fract. Mech.* **2022**, *259*, 108159. [CrossRef]
20. Aliha, M.R.M.; Samareh-Mousavi, S.S.; Mirsayar, M.M. Loading rate effect on mixed mode I/II brittle fracture behavior of PMMA using inclined cracked SBB specimen. *Int. J. Solids Struct.* **2021**, *232*, 111177. [CrossRef]
21. Feng, G.; Wang, X.; Kang, Y.; Zhang, Z. Effect of thermal cycling-dependent cracks on physical and mechanical properties of granite for enhanced geothermal system. *Int. J. Rock Mech. Min.* **2020**, *134*, 104476. [CrossRef]
22. Feng, G.; Wang, X.; Wang, M.; Kang, Y. Experimental investigation of thermal cycling effect on fracture characteristics of granite in a geothermal-energy reservoir. *Eng. Fract. Mech.* **2020**, *235*, 107180. [CrossRef]
23. Tang, S. The effect of T-stress on the fracture of brittle rock under compression. *Int. J. Rock Mech. Min.* **2015**, *79*, 86–98. [CrossRef]
24. Tang, S.; Bao, C.; Liu, H. Brittle fracture of rock under combined tensile and compressive loading conditions. *Can. Geotech. J.* **2017**, *54*, 88–101. [CrossRef]
25. Tang, S.; Zhang, H. The characteristics of crack initiation at the crack tip in rock material. *Chin. J. Rock. Mech. Eng.* **2017**, *36*, 552–561. (In Chinese)
26. Wu, S.; Li, L.; Zhang, X. *Rock Mechanics*; Higher Educational Press: Beijing, China, 2021; pp. 28–29.
27. Ma, F.; Liu, G.; Zhao, Z.; Xu, H.; Wang, G. Coupled Thermo-Hydro-Mechanical Modeling on the Rongcheng Geothermal Field, China. *Rock Mech. Rock Eng.* **2022**, *55*, 5209–5233. [CrossRef]
28. Li, T.; Tang, C.; Rutqvist, J.; Hu, M. TOUGH-RFPA: Coupled thermal-hydraulic-mechanical Rock Failure Process Analysis with application to deep geothermal wells. *Int. J. Rock Mech. Min.* **2021**, *142*, 104726. [CrossRef]

29. Wang, K.; Liu, Z.; Zeng, T.; Wang, F.; Shen, W.; Shao, J. Performance of enhanced geothermal system with varying injection-production parameters and reservoir properties. *Appl. Therm. Eng.* **2022**, *207*, 118160. [CrossRef]

30. Zhang, X.; Kong, B.; Yu, S.; Zhai, J.; Zhang, W.; Wang, R. Time series characteristics and difference analysis of AE of coal deformation and fracture under different damage and failure conditions. In Proceedings of the Third International Conference on Optoelectronic Science and Materials (ICOSM 2021), Hefei, China, 10–12 September 2021.

31. Zhang, C.; Yu, B.; Feng, Y.; Yuan, J.; Deng, J. Integrated Approach for the Wellbore Instability Analysis of a High-pressure, High-temperature Field in the South China Sea. In Proceedings of the IOP Conference Series: Earth and Environmental Science, Beijing, China, 23–26 October 2020.

32. Wang, C.; Liu, Z.; Zhou, H.; Wang, K.; Shen, W. A novel true triaxial test device with a high-temperature module for thermal-mechanical property characterization of hard rocks. *Eur. J. Environ. Civ. Eng.* **2022**. [CrossRef]

33. Guo, T.; Tang, S.; Liu, S.; Liu, X.; Zhang, W.; Qu, G. Numerical simulation of hydraulic fracturing of hot dry rock under thermal stress. *Eng. Fract. Mech.* **2021**, *240*, 107350. [CrossRef]

34. Kumari, W.G.P.; Ranjith, P.G.; Perera, M.S.A.; Li, X.; Li, L.H.; Chen, B.K.; Avanthi Isaka, B.L.; De Silva, V.R.S. Hydraulic fracturing under high temperature and pressure conditions with micro CT applications: Geothermal energy from hot dry rocks. *Fuel* **2018**, *230*, 138–154. [CrossRef]

35. Meng, Q.; Liu, J.; Pu, H.; Yu, L.; Wu, J.; Wang, C. Mechanical properties of limestone after high-temperature treatment under triaxial cyclic loading and unloading conditions. *Rock Mech. Rock Eng.* **2021**, *54*, 6413–6437. [CrossRef]

36. Ding, K.; Wang, L.; Ren, B.; Li, Z.; Wang, S.; Jiang, C. Experimental study on relative permeability characteristics for CO_2 in sandstone under high temperature and overburden pressure. *Minerals* **2021**, *11*, 956. [CrossRef]

37. Ergodan, F.; Sih, G.C. On the crack extension in plates under plane loading and transverse shear. *J. Basic Eng.* **1963**, *85*, 519–525.

38. Mirsayar, M.M.; Razmi, A.; Aliha, M.R.M.; Berto, F. Frictional crack initiation and propagation in rocks in rock materials. *Eng. Fract. Mech.* **2018**, *190*, 186–197. [CrossRef]

39. Williams, M.L. On the stress distribution at the base of a stationary crack. *ASME. J. Appl. Mech.* **1957**, *24*, 109–114. [CrossRef]

40. Saghafi, H.; Ayatollahi, M.R.; Sistaninia, M. A modified MTS criterion (MMTS) for mixed-mode fracture toughness assessment of brittle materials. *Mater. Sci. Eng. A-Struct. Mater. Prop. Microstruct. Process.* **2010**, *527*, 5624–5630. [CrossRef]

41. Ayatollahi, M.R.; Aliha, M.R.M. Wide range data for crack tip parameters in two disc-type specimens under Mixed-mode loading. *Comput. Mater. Sci.* **2017**, *38*, 660–670. [CrossRef]

42. Yang, J.; Lian, H.; Liang, W.; Nguyen, V.P.; Chen, Y. Experimental investigation of the effects of supercritical carbon dioxide on fracture toughness of bituminous coals. *Int. J. Rock Mech. Min.* **2018**, *107*, 233–242. [CrossRef]

43. Wei, M.; Dai, F.; Xu, N.; Liu, Y.; Zhao, T. Fracture prediction of rocks under mode I and mode II loading using the generalized maximum tangential strain criterion. *Eng. Fract. Mech.* **2017**, *186*, 21–38. [CrossRef]

44. Aliha, M.R.M.; Ayatollahi, M.R. Two-parameter fracture analysis of SCB rock specimen under mixed mode loading. *Eng. Fract. Mech.* **2013**, *103*, 115–123. [CrossRef]

45. Schmidt, R.A. A microcrack model and its significance to hydraulic fracturing and fracture toughness testing. In Proceedings of the 21st US Rock Mechanics Symposium, Rolla, MI, USA, 27–30 May 1980; pp. 581–590.

46. Alneasan, M.; Behnia, M.; Bagherpour, R. Frictional crack initiation and propagation in rocks under compressive loading. *Theor. Appl. Fract. Mech.* **2018**, *97*, 189–203. [CrossRef]

47. Meng, T.; Xue, G.; Ma, J.; Yang, Y.; Liu, W.; Zhang, J.; Jiao, B.; Fang, S.; Ren, G. Mixed mode fracture tests and inversion of FPZ at crack tip of overlying strata in underground coal gasification combustion cavity under real-time high temperature condition. *Eng. Fract. Mech.* **2020**, *239*, 107298. [CrossRef]

48. Feng, G.; Liu, C.; Wang, J.; Tao, Y.; Duan, Z.; Xiang, W. Experimental study on the fracture toughness of granite affected by coupled mechanical-thermo. *Lithosphere* **2022**, *2022*, 5715093. [CrossRef]

49. Kuruppu, M.D.; Obara, Y.; Ayatollahi, M.R.; Chong, K.P.; Funatsu, T. ISRM suggested method for determining the Mode I fracture toughness using semi-circular bend specimen. *Rock Mech. Rock Eng.* **2014**, *47*, 267–274. [CrossRef]

50. Sun, Z.; Feng, G.; Song, X.; Meng, T.; Zhu, D.; Zhai, Y.; Wang, Z. Effects of CO_2 state and anisotropy on the progressive failure characteristics of bituminous coal: An experimental study. *Chin. J. Rock. Mech. Eng.* **2022**, *41*, 70–81. (In Chinese)

51. Suo, Y.; Chen, Z.; Rahman, S.S.; Song, H. Experimental and numerical investigation of the effect of bedding layer orientation on fracture toughness of shale rocks. *Rock Mech. Rock Eng.* **2020**, *53*, 3625–3635. [CrossRef]

52. Meng, F.; Song, J.; Wong, L.N.Y.; Wang, Z.; Zhang, C. Characterization of roughness and shear behavior of thermally treated granite fractures. *Eng. Geol.* **2021**, *293*, 106287. [CrossRef]

53. Zhao, F.; Sun, Q.; Zhang, W. Environ. Fractal analysis of pore structure of granite after variable thermal cycles. *Earth Sci.* **2019**, *78*, 677. [CrossRef]

54. Xiao, W.; Yu, G.; Li, H.; Zhang, D.; Li, S.; Yu, B. Thermal cracking characteristics and mechanism of sandstone after high-temperature treatment. *Fatigue Fract. Eng. Mater. Struct.* **2021**, *44*, 3169–3185. [CrossRef]

55. Taheri, A.; Zhang, Y.; Munoz, H. Performance of rock crack stress thresholds determination criteria and investigating strength and confining pressure effects. *Constr. Build. Mater.* **2020**, *243*, 118263. [CrossRef]

56. Feng, G.; Kang, Y.; Chen, F.; Liu, Y.; Wang, X. The influence of temperatures on mixed-mode (I plus II) and mode-II fracture toughness of sandstone. *Eng. Fract. Mech.* **2018**, *189*, 51–63. [CrossRef]

57. Meng, Q.; Qian, W.; Liu, J.; Zhang, M.; Lu, M.; Wu, Y. Analysis of triaxial compression deformation and strength characteristics of limestone after high temperature. *Arab. J. Geosci.* **2020**, *13*, 153. [CrossRef]
58. Funatsu, T.; Kuruppu, M.; Matsui, K. Effects of temperature and confining pressure on mixed-mode (I-II) and mode-II fracture toughness of Kimachi sandstone. *Int. J. Rock Mech. Min.* **2014**, *67*, 1–8. [CrossRef]
59. Li, Y.; Dai, F.; Wei, M.; Du, H. Numerical investigation on dynamic fracture behavior of cracked rocks under mixed mode I/II loading. *Eng. Fract. Mech.* **2020**, *235*, 107176. [CrossRef]

Article

Study of the Characterisation Method of Effective Two-Phase Seepage Flow in the Construction of Gas Storage Reservoirs

Guoying Jiao [1], Shijie Zhu [1,*], Fei Xie [2], Shuhe Yang [2], Zuping Xiang [1] and Jiangen Xu [1]

[1] School of Petroleum and Natural Gas Engineering, Chongqing University of Science & Technology, Chongqing 401331, China
[2] Research Institute of Exploration and Development, Dagang Oilfield Company, PetroChina, Tianjing 300450, China
*   Correspondence: zhusj@cqust.edu.cn

Abstract: During the rebuilding of a gas reservoir, repeated "strong injection and mining" processes change the seepage capacities of gas and water. Hence, accurately determining the seepage laws of gas and water in a gas storage reservoir is crucial. In this study, a standard relative permeability test was conducted with a one-dimensional core. Additionally, a gas reservoir injection and mining simulation experiment was conducted with a two-dimensional plate. The results show that the relative permeability curve obtained by the one-dimensional core test did not accurately reflect the operation characteristics of the gas storage and the change in the seepage law during the gas reservoir construction. Furthermore, in the two-dimensional plate experiment, the operation mode was restored using the plane radial flow formula, the mutual relationship between the gas and water's effective permeability under different injection stages was established, and the multi-cycle injection operation was accurately described. This method lays the foundation for the construction of gas reservoirs and the establishment of the multi-phase seepage law for gas reservoirs.

Keywords: gas storage; injection production simulation; effective permeability; seepage law; two-dimensional simulation experiment

Citation: Jiao, G.; Zhu, S.; Xie, F.; Yang, S.; Xiang, Z.; Xu, J. Study of the Characterisation Method of Effective Two-Phase Seepage Flow in the Construction of Gas Storage Reservoirs. *Energies* 2023, 16, 242. https://doi.org/10.3390/en16010242

Academic Editor: Hossein Hamidi

Received: 10 November 2022
Revised: 14 December 2022
Accepted: 19 December 2022
Published: 26 December 2022

1. Introduction

The purpose of gas storage is to solve the inherent contradictions between natural gas supply and consumption [1], that is, the inherent contradictions between reliable, safe, stable, and continuous gas supply and consumption demand imbalance [2]. With the increasing proportion of natural gas in the energy structure [3], countries around the world are developing more gas storage reservoirs to increase gas storage [4], thereby ensuring peak emergency regulation [5]. To meet the construction needs of gas storage reservoirs [6], depleted gas reservoirs are gradually reconstructed. Meanwhile, the "strong injection and mining" operation in the construction process of gas storage reservoirs significantly affects the throat structure [7] and seepage capacity of the reservoir [8], as well as the size of the gas storage library building space [9] and the long-term operation safety [10]. Therefore, it is particularly important to accurately determine the relative permeability curve rules in the strong injection and mining construction process of gas storage reservoirs [11].

The conventional method for studying the law of fluid seepage is based on relative permeability tests [12]. Hence, to simulate the operation of gas storage reservoirs, experiments were conducted to establish the relationship between the injection–production process and the reverse displacement to determine the relative permeability curves of gas/water, oil/water, and gas/oil. Meanwhile, the injection–production operation of gas storage reservoirs was achieved through repeated displacements. Considering the relative permeability test of oil/water as an example, the specific experimental steps were as follows [13]: (1) The water was first saturated, and then the oil was saturated (oil displaced the water to the bound state). (2) The oil phase permeability in the bound water state was determined. (3) Water drive oil experiments were performed according to the displacement conditions, and a suitable displacement speed or displacement pressure difference is selected. (4) The direction of the rock sample switched from left to right and they were loaded into the pressure-tapped core based on the displacement conditions. An appropriate displacement speed or pressure difference was selected for the oil/water displacement experiment. (5) The rock sample changed direction from left to right in the pressure-tapped core, and this was repeated to perform multiple relative permeability curve tests for the mutual oil/water drives. In the present study, the underwater phase seepage deviation of the last two rounds of water drive oil binding was within 3% at the end of the oil/water mutual drive experiment. (6) The "J·B·N" method was applied for data processing, and the relative permeability curves were obtained, as shown in Figure 1 [14].

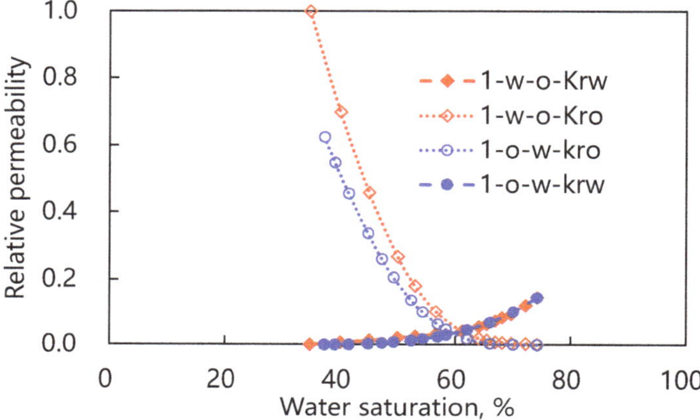

Figure 1. Oil/water permeability plot of relative mutual drive.

However, the seepage law in the construction process of gas storage reservoirs cannot be accurately reflected based on the above process because of the following reasons: (1) There are differences in the injection mining and mutual drive methods. The displacement process in the relative permeability test caused steady seepage at a constant speed or pressure. However, the injection and mining process caused unstable seepage at a constant pressure. (2) The relative permeability test simulated the whole process of the reservoir construction (as shown in Figure 1), ranging from bound water saturation to residual oil saturation and vice versa. However, one gas storage injection and mining operation cycle is a very small stage of the gas storage operation. (3) The core ripple efficiency in the relative permeability test was 100%, whereas the injection efficiency in the actual injection process was considerably less than 100%, indicating a large difference in the seepage law. Therefore, to study the seepage law of the gas reservoir, it was necessary to learn from the high-accurate simulation of the appropriate equipment model, adopt a more appropriate method, and establish a more accurate seepage law to lay the foundation for the numerical simulation of the gas reservoir [15].

The scheme optimisation during the construction of an existing gas storage reservoir depends on the reservoir numerical simulation technology, and the core technology is a full-cycle simulation of the injection and production process using the relative permeability data. The accuracy of the relative permeability curve will directly affect the understanding of the simulation results, so it is particularly important to obtain a highly matched relative permeability curve of the gas storage reservoir. Moreover, there is no relevant literature reporting on a new method for measuring the relative permeability curve during the construction of a gas storage reservoir. Thus, considering the influence of sweep efficiency in the injection–production process, this study was conducted using a two-dimensional plane model. First, the change in the seepage law after multiple rounds of the injection–production process (mutual drive) was determined. Second, combined with the plane radial flow theory, the change in the effective permeability of the gas storage reservoir during the injection–production process was studied. Through the relative permeability curve, which is the essential attribute of the relationship between water saturation and permeability, a highly simulated experiment was designed to analyse the characteristics of the relationship between water saturation and permeability, and a method for analysing the two-phase seepage ability of gas storage reservoirs during the injection–production process was established using a two-dimensional experimental model.

2. Experimental Methods

The experiments in this study were divided into a one-dimensional core experiment and a two-dimensional core experiment. The one-dimensional core test was conducted according to the industry standard of a relative permeability curve. A multi-cycle injection and production process of gas storage was simulated using multi-cycle mutual drive [16]. This is the main method for obtaining the multi-cycle injection production relative permeability at this stage. The two-dimensional core experiment was a complete experimental process formed by reducing the injection and mining process of the ore field using the real plane model. Through the description and comparison of the two experimental processes, and combined with the presentation of data, this study provides support for the establishment of new methods.

2.1. Experimental Conditions

Based on the properties of a gas reservoir, a one-dimensional artificial core and a two-dimensional plate were adopted in this study. The specific parameters were as follows:

(1) Experimental water: The viscosity of the formation water was 0.894 mPa·s;

(2) Natural gas: Methane gas was used in the experiment, with a viscosity of 0.0178 mPa·s. Its components are listed in Table 1.

Table 1. Methane gas fractions.

Component	C_1	C_2	C_3	C_4	C_5	N_2	CO_2	He	H_2
Content	91.41	4.93	0.96	0.41	0.24	1.63	0.06	0.29	0.07

(3) Experimental core: An artificial core comprising quartz sand, solid epoxy resin with an average permeability of 1000 mD, and a flat plate model with a size of $500 \times 500 \times 20$ mm was adopted;

(4) Experimental temperature: The simulated reservoir temperature was 82 °C;

(5) Experimental device: The device comprised a one-dimensional artificial core with a permeability of approximately 800 mD and a two-dimensional plate model, as shown in Figure 2. The inner diameter of the device was $500 \times 500 \times 20$ mm, and the upper limit pressure was 20 MPa.

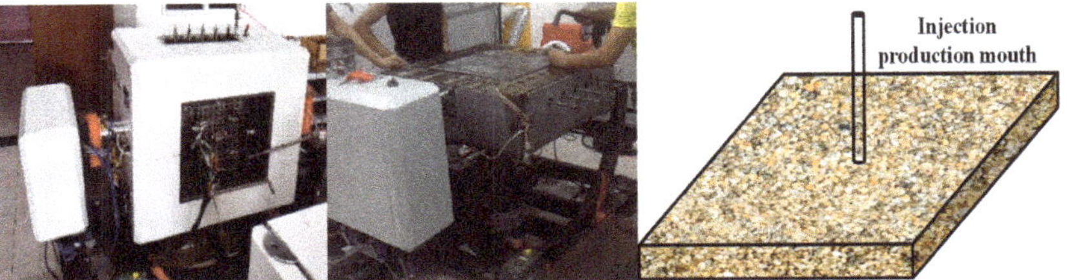

Figure 2. Two-dimensional core simulation setup: (**left**) overall appearance; (**middle**) plate placement, (**right**) schematic diagram of the injection and mining method.

2.2. Experimental Procedure

2.2.1. Experimental Procedure of the One-Dimensional Core

The relative permeability test was performed according to industry standards, and the steps are given as follows [17]: (1) The core was dried, and the gas measurements were taken by injecting gas into the core. The permeability of the core was then determined as the absolute permeability. (2) The pore volume/porosity of the core was determined by saturating the rock sample with formation water. (3) The water logging permeability (or absolute permeability) of the rock sample was measured after injecting the formation water into the sample at a constant speed. (4) The gas flooding required that the initial differential pressure could overcome the end effect without turbulence. (5) The displacement pressure difference, cumulative fluid production, cumulative water production, and initial gas breakthrough point were recorded at each displacement time. When the gas drive water reached the residual water state, the gas permeability was measured, and the experiment was terminated.

Based on the one-dimensional two-phase seepage theory and gas state equation, the gas flooding experiment of the rock samples was conducted using the unsteady constant pressure method [18]. The gas production, water production, and differential pressure at both ends of the rock sample at each time point at the outlet of the rock sample during the gas flooding were recorded. Moreover, the J·B·N method (Equations (1)–(8)) was used to calculate the gas/water relative permeability and the corresponding water saturation of the rock samples. Afterwards, the gas/water relative permeability curve was obtained [19].

$$S_{gav} = \frac{V_w}{V_p} \tag{1}$$

$$\frac{K_{rg}}{K_{rw}} = \frac{f_g}{f_w} \times \frac{\mu_g}{\mu_w} \tag{2}$$

$$q_{gi} = \frac{\Delta V_{gi}}{\Delta t} \tag{3}$$

$$f_g = \frac{\Delta V_{gi}}{\Delta V_i} \tag{4}$$

$$K_{rg} = \frac{q_{gi}}{q_g} \tag{5}$$

$$C = \frac{p_1}{p_2 + \Delta p} \tag{6}$$

$$q_g = \frac{KA}{\mu_g L} \times \Delta p \tag{7}$$

$$f_w = \frac{\Delta V_{wi}}{\Delta V_i} \tag{8}$$

where S_{gav} is the average gas saturation (%), V_w is the cumulative outlet water volume (mL); V_p is the cumulative export gas volume (mL); K_{rg} is the relative permeability of the gas phase (in decimals); K_{rw} is the relative permeability of the water phase (in decimals); f_g is the gas cut (in decimals); f_w is the water cut (in decimals); μ_g is the injection gas viscosity (mPa·s); μ_w is the viscosity of the simulated formation water in the saturated rock samples (mPa·s); q_{gi} is the gas flow rate during the two-phase flow (mL/s); V_{gi} is the gas increment measured at the outlet pressure (mL); V_i is the amount of fluid change at a certain time interval (mL); q_g is the gas flow during the single-phase flow (mL/s); C is the antihypertensive volume factor (in decimals); p_1 is the absolute inlet pressure of the rock sample (MPa); p_2 is the absolute outlet pressure of the rock sample (MPa); and V_{wi} is the measured water increment at a certain time interval (mL).

2.2.2. Experimental Procedure of the Two-Dimensional Model

(1) The two-dimensional plate model was filled with quartz sand and then properly sealed. The "one note, four minings" mode of saturated formation water and the "four injections, one mining" mode of saturated crude oil were used. The formation water and crude oil were injected at a constant speed (1 mL/min), and the pressure was maintained at 10 MPa;

(2) A multi-cycle injection test was conducted using the following steps: (i) The injection and mining pressure of the actual simulation experiment was 10–17 MPa. (ii) One mining well network was formed at the centre of the model. (iii) In the injection–production experiment, within the upper- and lower-limit operating pressure ranges of the simulated gas storage, the injected gas was gradually pressurised to the target constant pressure, and the injection time (simulated mine injection time) was recorded. Afterwards, a boiling operation was simulated for 1/10 of the injection time, and the exhaust valve was opened. Meanwhile, the time (half of the injection time) was controlled until the pressure at the outlet reached the atmospheric pressure and no water was produced. Step (iii) was then repeated five times, and the output volume was recorded and measured using a gas flow meter. (iv) The injection formed the central well, and the model was injected in the middle and extracted in the middle. Figure 2 shows the experimental setup.

3. Results and Discussion

The injection production operation mode in the process of gas storage reservoir construction determines the multi-cycle mutual displacement of gas and water. Therefore, the multi-cycle mutual drive mode was introduced for characterisation based on the relative permeability. In addition, a two-dimensional large plane model was built to conduct multi-cycle injection and production research in combination with actual injection and production parameters, and the operation process of database construction was highly restored. A comparative study of the two methods was expected to establish a highly consistent pattern of the law of infiltration.

3.1. Two-Phase Percolation Experiment of the One-Dimensional Core

The relative permeability curves of the gas/water mixture were obtained, as shown in Figure 3.

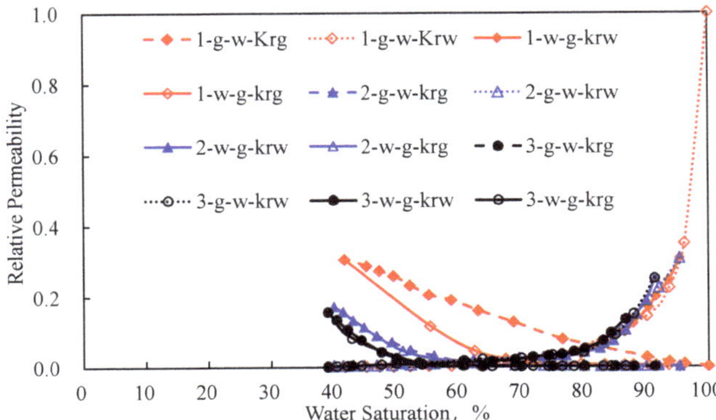

Figure 3. Relative permeability curves of gas/water in the highly permeable core.

Based on the relative permeability curve under the target core condition, the following conclusions can be drawn: (1) The seepage capacities of the gas drive water and water drive gas were based on two laws, which were caused by the wetting characteristics of the core; hence, the simulation resulted in an inaccurate with the standard relative permeability curve compared to the experimental means. (2) After multiple gas/water mutual drives, the residual gas saturation and bound water saturation changed, indicating that the effective storage capacity after multiple rounds was reduced and not conducive to an increase in the gas storage space. The large decrease in the seepage capacity indicates a decrease in gas storage at the later stage of secondary injection mining; however, this is obviously different from the actual situation characteristics of the gas reservoir, which showed that the seepage curve of the one-dimensional core and gas storage simulation was limited.

3.2. Two-Phase Seepage Experiment of the Two-Dimensional Plate

3.2.1. Experimental Result

(1) The gas injection time, the shut-in time, the oil return time during the simulation process, and the experimental data of the water and gas productions are listed in Table 2, and the single-cycle and cumulative extraction degrees are shown in Figure 4.

Table 2. Gas and water production under different injection and mining cycles.

Construction/Water Characteristics	Injection–Production Cycle								Remarks
	One	Two	Three	Four	Five	Six	Seven	Eight	
Gas injection time (min)	25	23	20	19	15	15	15	10	
Well shut-in time (min)	2.5	2.5	2	2	1.5	1.5	1.5	1	The saturated water volume was 2010 mL
Oil return time (min)	10	11	10	10	7	7	7	5	
Water production (mL)	302	206	131	97	88	65	41	23	953
Gas production (L)	7.2	19.4	25.9	36.4	46.1	51.8	57.5	64.8	309.1

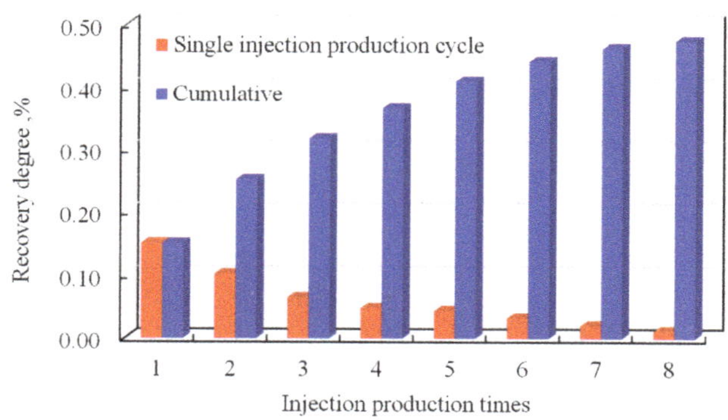

Figure 4. Extraction degrees under different injection cycles.

As shown in Figure 4 and Table 2, during the injection and mining operations of the gas storage tank, the water could be collected by extracting the reservoir fluid to excavate more gas storage space. With the progress of the injection and mining cycles, the water extraction degree in a single cycle gradually decreased, whereas the cumulative extraction degree gradually increased and stabilised until no more water was produced, thereby achieving the bound water conditions, that is, the maximum effective storage space was reached. The analysis indicates that after the multi-cycle injection and extraction, the remaining fluid of the reservoir decreased, and the aqueous seepage capacity was significantly reduced; thus, the extraction degree gradually decreased. The large increase in gas output was because of the large increase in the gas storage space, causing more gas to be injected under the upper-limit pressure of the gas injection.

It can be deduced from Figure 4 that the water saturation decreased with the injection cycle, and the production velocity per unit of time indicates that the permeability of the effective gas injection space in the model and the gas output (gas saturation) increased.

This law is similar to the two-phase seepage law expressed in the gas/water relative permeability curve, indicating that the multi-cycle injection and mining process can also be used to build a similar relative permeability curve, which is the seepage law most consistent with the injection and mining process.

3.2.2. Method Principle

Considering the two-dimensional plate model, the plane radial flow formula was adopted to calculate the seepage flow capacity, which is important for predicting and determining the flow capacity of a gas well.

One of the most important scenarios for applying the planar radial flow formula is the determination of the effective permeability of the reservoir. Based on the relevant literature [20,21], the planar radial flow formula is given by:

$$\frac{P_e - P_{wf}}{m_t} = \frac{\lambda^n}{2\pi Kh} m_t + \frac{1}{2\pi Kh}\left(\ln \frac{r_e}{r_w} + S\right) \tag{9}$$

where P_e is the reservoir supply pressure (MPa); P_{wf} is the flow pressure at the well bottom (MPa); m_t is the mass flow (kg/d); λ is the turbulent effect coefficient ($\lambda \geq 1$; $\lambda = 1$ is the laminar flow); n is the representation index of high and low speeds; K is the effective permeability of the reservoir (mD); h is the effective thickness of the reservoir (m); r_e is the water supply radius (m); r_w is the Holbore radius (m); and S is the epidermal coefficient.

In Equation (9), the size of the model was determined to be $500 \times 500 \times 20$ mm, the injection well was in the centre of the model, and the experimental research pipeline was a 2 mm capillary. Additionally, the oil discharge radius, wellbore radius, and reservoir thickness were determined to be 250, 1, and 20 mm, respectively. Furthermore, both the formation water and natural gas were determined based on the basic experimental parameters, and the mass flow was determined from the results of the two-dimensional plane radial flow. Subsequently, the Reynolds equation was applied to determine the laminar flow below 3 MPa and the turbulence above 3 MPa, depending on the flow velocity changes resulting from different pressures. The application of the n was then determined, and the turbulence coefficient was obtained using empirical values.

3.2.3. Method Application Examples

The steps for determining the changes in the effective permeability during the gas storage injection were as follows: (1) Based on the physical conditions of the target block, a two-dimensional large-scale plane rock plate model was constructed, and the relevant parameters (oil/water physical properties, pressure, temperature, and so on) were determined. (2) Multiple rounds of gas injection and production tests were performed according to the injection and production operation scheme of the gas storage in the target reservoir. (3) The plane radial flow formula (Equation (9)) was used to calculate the effective permeability based on the changes in water/gas production. (4) The water saturation was calculated according to the recovery degree, and the relative permeability curve was constructed in combination with the calculated effective permeability.

(2) Equation (9) was used to calculate the mean core permeability at the current cycle from the actual experimental model in 2D (see Table 3).

Table 3. Permeability values for the calculation of the two-dimensional model.

Period	1	2	3	4	5	6	7	8
Mean water-phase permeability (mD)	990	870	740	540	490	360	230	130
Mean gas-phase permeability (mD)	120	240	320	450	570	640	710	800

As presented in Table 3, the permeability values reflect the average permeability characteristics of the core after an injection and mining cycle, which are the equilibrium values after its strong injection and mining. This reflects the impact of the injection and mining process in which the water content saturation and the water-phase permeability capacity gradually decreased, whereas the gas-phase permeability capacity gradually increased. The experimental data present the key parameters of the permeability curve. Therefore, the saturation and fluid volume data under each cycle were applied to build a relationship between the permeability and saturation, and the data were combined with the absolute permeability measured by the core to calculate the relative permeability and the change in the relative permeability (see Figure 5).

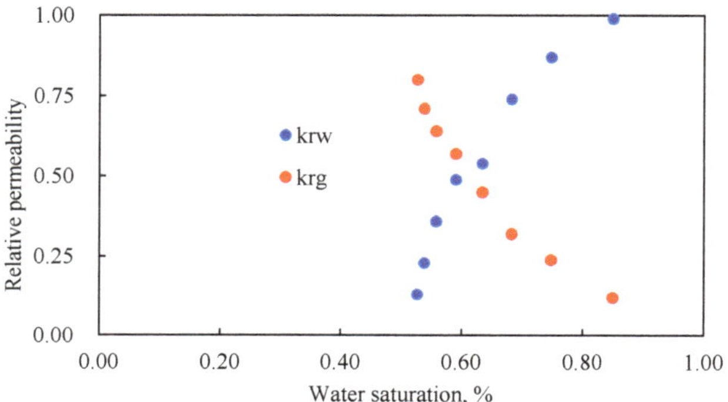

Figure 5. Relative permeability plot.

The relative permeability curve features shown in Figure 5 are almost consistent with those of the one-dimensional core test; however, the significance of the seepage law is different. The relative permeability curve was a relatively stable seepage capacity value after the mutual injection and mining process. This reflects the influence of the injection and mining process on the seepage law of gas storage through continuous multi-cycle gas injections.

3.3. Flow Difference between the One-Dimensional Core and Two-Dimensional Plate

Figures 3 and 5 show the relative permeability curves under the two experimental conditions. In fact, the relative permeability curves themselves are the simulated features of a reservoir and gas reservoir from the original state to extreme development, or the simulated process of oil and gas reservoir formation. Although the simulation of the injection–production process was repeated several times, the results do not accurately describe the injection and production of gas storage. Therefore, the set of curves adopted in the multiple rounds of the mutual drive process is not accurate.

The seepage curve constructed by the injection simulation experiment of the two-dimensional rock plate represents the final seepage law of the gas reservoir under each injection and mining cycle and the entire construction process of the simulated gas storage reservoir. Based on the data of the one-dimensional permeability experiment, the relative permeability corresponding to the water content saturation after the two-dimensional injection and mining experiment is given by Equation (1). The calculation results were compared with the actual data, as presented in Table 4.

Table 4. Multi-cycle injection and production results calculated from the relative permeability curve of the one-dimensional core.

Period	1	2	3	4	5	6	7	8
Water yield calculated for the first secondary phase seepage (mL)	217	153	112	87	74	54	32	5
Water yield calculated for the third round of the secondary phase seepage (mL)	117	81	63	43	18	14	9	2
Actual fluid yield (mL)	302	206	131	97	88	65	41	23

The relative permeability curve obtained from the corresponding water saturation in the injection and mining core was significantly lower than that of the actual experimental data. Based on the analysis, the two-dimensional core model simulated the injection process of gas storage, and the strong injection and mining production mode significantly enhanced the seepage capacity of the fluid in porous media and relatively increased its effective permeability. Meanwhile, the water-phase permeability capacity after the third mutual drive in the phase–seepage curve significantly decreased, with a more significant decrease in the water-production capacity. Therefore, the relative permeability curve obtained by the one-dimensional core does not accurately reflect the seepage law of the gas storage injection operation, whereas the two-dimensional plate experiment more accurately reflects the seepage law of the gas storage injection operation time.

The relative permeability curve obtained by the two-dimensional experimental model was analysed and fitted during the process of the reservoir numerical simulation. The results are closer to the field data than the one-dimensional relative permeability experimental curve. This lays a foundation for accurately guiding the development of field tests.

4. Conclusions

(1) During multiple gas/water mutual drive rounds, the seepage capacity of the core flow significantly decreased. Thus, the permeability capacity of a multi-cycle injection operation could not be scaled with a single relative permeability curve.

(2) The two-dimensional simulation experiment accurately simulated the operation process of gas storage, and gas injection improved the storage capacity. With each injection cycle, the synergistic water extraction capacity gradually decreased. Moreover, the relative permeability curve obtained by the two-dimensional simulation model experiment and the plane radial flow formula was the seepage law after the multi-cycle injection and mining process. Hence, the phase–seepage relationship derived from the two-dimensional simulation experiment is more consistent with the results of the gas storage injection operation.

Author Contributions: Methodology, S.Z.; Software, S.Y.; Formal analysis, G.J.; Resources, F.X.; Writing—original draft, G.J. and S.Z.; Writing—review & editing, Z.X.; Funding acquisition, S.Z. and J.X. All authors have read and agreed to the published version of the manuscript.

Funding: This work was funded by the Natural Science Foundation of Chongqing (cstc2021jcyj-msxmX0522 and cstc2020jcyj-msxmX0163), the Science and Technology Research Program of Chongqing Municipal Education Commission (KJQN202001518), and the Scientific Research Funding Project of Chongqing University of Science and Technology (ckrc2021004).

Data Availability Statement: The data supporting the findings of this study are available from the corresponding author upon reasonable request.

Conflicts of Interest: The authors declare no conflict of interest.

References

1. Zheng, D.; Xu, H.; Wang, J.; Sun, J.; Zhao, K.; Li, C.; Shi, L.; Tang, L. Key evaluation techniques in the process of gas reservoir being converted into underground gas storage. *Pet. Explor. Dev.* **2017**, *44*, 794–801. [CrossRef]
2. Zeng, D.; Zhang, J.; Zhang, G.; Mi, L. Research progress of Sinopec's key underground gas storage construction technologies. *Nat. Gas Ind.* **2020**, *40*, 115–123.
3. Ma, X.; Ding, G. *China Underground Natural Gas Storage*; Petroleum Industry Press: Beijing, China, 2018.
4. Tuna, E.; Can, P. Natural gas underground storage and oil recovery with horizontal wells. *J. Pet. Sci. Eng.* **2020**, *187*, 106753.
5. Liu, W.; Jiang, D.; Chen, J.; Daemen, J.J.K.; Tang, K.; Wu, F. Comprehensive feasibility study of two-well-horizontal caverns for natural gas storage in thinly-bedded salt rocks in China. *Energy* **2017**, *143*, 1006–1019. [CrossRef]
6. Zeng, S. *Study on Seepage Mechanism and Application of Gas Storage Reconstruction in Late High Water Cut Reservoir*; Southwest Petroleum Institute: Chengdu, China, 2005.
7. He, S.; Men, C.; Zhou, J.; Wu, Z. Study on injection production seepage characteristics of Da zhang tuo gas reservoir. *Nat. Gas Ind.* **2006**, *5*, 90–92.
8. Ban, F.; Gao, S.; Wang, J. Gas Injection-production Mechanism of Gas Storage in Depleted Oil Reservoirs. *Nat. Gas Geosci.* **2009**, *20*, 1005–1008.
9. Wang, J.; Guo, P.; Jiang, F. The Physical Simulation Study on the Gas-Drive Multiphase flow Mechanism of Aquifer Gas Storage. *Nat. Gas Geosci.* **2006**, *4*, 597–600.
10. Lei, S.; Guangzhi, L.; Wei, X.; Shusheng, G.; Tong, G. Gas-water percolation mechanism in an underground storage built on a water-drive sandstone gas reservoir. *Nat. Gas Ind.* **2012**, *32*, 85–87.
11. He, X.; Huang, S.; Sun, C.; Xu, J.; Zhang, X.; Yang, X. Gas-water relative permeability variation in multi-cycle injection production of underground gas storage in flooded depleted gas reservoir. *Oil Gas Storage Transp.* **2015**, *34*, 150–153.
12. Shi, L.; Wang, J.; Zhu, H.; Duan, Y. Microscopic percolation laws of gas and water in underground gas storages rebuilt from water-invasion gas reservoirs. *Nat. Gas Explor. Dev.* **2020**, *43*, 58–63.
13. Ding, Y.; Zhang, Q.; Shi, L.; Sun, j.; Jiang, S. Characteristics of Underground Gas Storage Built on Carbonate Gas Reservoir in S Region, North China. *Sci. Technol. Guide* **2014**, *32*, 51–54.
14. Sun, C.; Hou, J.; Shi, L.; Liang, C.; Gan, Z. A physical simulation experimental system for injection-withdrawal operation of gas reservoir underground gas storage and its application. *Nat. Gas Ind.* **2016**, *36*, 58–61.
15. Niu, C. *Seepage Displacement and Operating Simulation of Fractured Underground Natural Gas Storage Reservoir*; Harbin Institute of technology: Harbin, China, 2016.
16. Zhu, S. *Study on the Timing of Polymer Flooding in Common Heavy Oil Reservoir*; Southwest Petroleum University: Chengdu, China, 2015.
17. Sinan, Z.H.U.; Junchang, S.U.N.; Guoqi, W.E.I.; Zheng, D.; Jieming, W.A.N.G.; Lei, S.H.I.; Xianshan, L.I.U. Numerical simulation-based correction of relative permeability hysteresis in water-invaded underground gas storage during multi-cycle injection and production. *Pet. Explor. Dev.* **2021**, *48*, 190–200.
18. Li, C.; Li, X.; Gao, S.; Liu, H.; You, S.; Fang, F.; Shen, W. Experiment on gas-water two-phase seepage and inflow performance curves of gas wells in carbonate reservoirs: A case study of Long wang miao Formation and Deng ying Formation in Gaoshiti-Moxi block, Sichuan Basin, SW China. *Pet. Explor. Dev.* **2017**, *44*, 930–938. [CrossRef]
19. Zhu, S.; Shi, L.; Zhang, J.; Li, T.; Wang, G.; Xue, X.; Ye, Z. Application method of relative permeability curve to judge the timing of polymer flooding to polymer injection. *Reserv. Eval. Dev.* **2020**, *10*, 128–134.
20. Zhai, Y. *Seepage Mechanics*; Petroleum Industry Press: Beijing, China, 2016.
21. Yao, J.; Gu, J.; Lu, A. *Principles and Methods of Reservoir Engineering*; China University of Petroleum Press: Dongying, China, 2016.

 *energies*

Article

Study on the Evolution Law of Deep Rock Cracks and the Mechanism of Graded Gradient Support

Zijie Hong [1], Zhenhua Li [2,*], Feng Du [2], Zhengzheng Cao [1] and Chun Zhu [3]

[1] School of Civil Engineering, Henan Polytechnic University, Jiaozuo 454003, China
[2] School of Energy and Engineering, Henan Polytechnic University, Jiaozuo 454003, China
[3] School of Earth Sciences and Engineering, Hohai University, Nanjing 210098, China
* Correspondence: hongzijie2010@126.com

Abstract: The surrounding rock of deep roadway is mostly composed of fractured rock. The deformation of roadway surrounding rock is complicated, which not only involves the stress change, but also involves the support means. This paper aims to study the deformation and fracture evolution law of surrounding rock in deep underground engineering. According to the stress rebalancing characteristics, after roadway excavation, the development and evolution characteristics of surrounding rock cracks are studied. At the same time, different seepage zones are divided according to the relationship between surrounding rock failure and its total stress–strain, that is, complete seepage zone, seepage shielding zone, and proto-rock seepage zone. The crack distribution characteristics of surrounding rock are studied, and the graded control of gradient support is proposed. In the broken area, the gradient bearing shell outside the roadway is achieved by means of bolting and high-strength grouting. As the cracks and pore sizes in the plastic zone gradually decrease along the radial stress direction of the roadway, and the open cracks gradually change into closed cracks, it is difficult for ordinary grouting materials to complete better consolidation and filling. Therefore, small particle size grouting reinforcement materials are studied. The plastic zone (fracture zone) is reinforced with nano-scale grouting material, and the internal three-dimensional gradient bearing shell is formed by combining with the anchor cable. This research plays an important guiding role in the stability of deep roadway surrounding rock.

Keywords: fracture evolution law; fractured seepage zone; gradient failure support; hierarchical control mechanism

Citation: Hong, Z.; Li, Z.; Du, F.; Cao, Z.; Zhu, C. Study on the Evolution Law of Deep Rock Cracks and the Mechanism of Graded Gradient Support. *Energies* **2023**, *16*, 1183. https://doi.org/10.3390/en16031183

Academic Editor: Mohammad Sarmadivaleh

Received: 12 December 2022
Revised: 4 January 2023
Accepted: 18 January 2023
Published: 20 January 2023

1. Introduction

With the continuous extension of coal mining depth, the underground mining environment becomes more complicated. After the excavation of deep roadways, the stress of the roadways will change significantly [1,2]. The tangential stress gradually weakens from the outside to the inside and reaches the maximum at the surface, shown as stress concentration. The radial stress increases from the outside to the inside, which is almost zero on the surface, resulting in layered failure of the surrounding rock, tangential force transfer, and further failure of the surrounding rock and crack development [3,4]. Under the influence of high-stress concentration and mining, the surrounding rock can show various cracks. For example, when the surrounding rock cracks develop and have a large opening degree, because of the stress weakening, a large number of cracks with a small opening degree can appear. Moreover, many of these cracks will be isolated and not connected with each other, resulting in poor permeability characteristics of the surrounding rock. The high development of the surrounding rock cracks is bound to reduce its strength, and with the continuous change of stress, it will cause large deformation of roadways and affect the safety of mining production.

Many scholars have studied the stability of roadway surrounding rock control in coal mines. At the same time, a variety of research methods have been used, including theoretical analysis, laboratory testing, etc., and achieved good results [5–7]. According to Li et al. [8], large-scale geomechanical models and stress evolution were used to study the deformation characteristics of deep roadway surrounding rock. Zhang et al. [9] studied the excavation and unloading of deep underground tunnels caused by the rupture of surrounding rock and analyzed the fracturing mechanism. Petuknov et al [10] discussed the deformation of surrounding rock after and the failure of stable rock. Alsayed [11] carried out experiments with rock cylinders as specimens and realized true triaxial tests, showing several different failure modes. Kulatilake et al. [12] studied the behavior of jointed rock mass and its failure summary standard. With the development of computer technology, numerical simulation is widely used in the research of coal mining [13,14]. The finite element method (FEM) and finite element difference method (FDM) were used to study roadway stability [15–17]. Yang et al. [18] studied the deformation and failure of surrounding rock by combining field investigation, laboratory experiment, and UDEC simulation. Research on the prevention and technology of traditional support, including bolt, anchor cable, and other means, is difficult to meet the control requirements, leading to the softening of coal and rock, resulting in continuous deformation and large-scale movement of surrounding rock [19,20]. Grouting reinforcement can change the shape of surrounding rock to a certain extent, and grouting can fill grout into surrounding rock cracks, improve its cohesion and strength, and further improve the stability of surrounding rock [21–23]. The good grouting effect of surrounding rock fissures can ensure that the broken surrounding rock is closely connected together, improve the overall performance of surrounding rock, cement the broken surrounding rock into a complete whole, and even restore to the original rock state to ensure its overall stability.

Under the circumstances of stress redistribution, the surrounding rock structure evolution occurs, which is manifested as obvious crack change from the outside to the inside of the roadway. That is, the cracks close to the roadway are dense and wide, and there are sparse and narrow cracks further inside. The different crack opening causes the traditional grouting method to be unscientific, and the grouting material particles are large, which greatly reduces its injectability and affects the grouting effect, which cannot meet the desired effect. Therefore, this paper studies the regularity and characteristics of the evolution of deep rock cracks through theoretical analysis and expounds the traditional support and reinforcement technology. At the same time, a new type of grouting material is developed, and combined with the characteristics of surrounding rock and cracks, the graded control mechanical model of gradient support is established. According to the gradient failure law, using the gradient control technology, the surrounding rock formed a three-dimensional gradient pressure shell, which restored the bearing capacity of the surrounding rock. The gradient model established in this paper provides a new idea for the control of deep surrounding rock. Meanwhile, the proposed hierarchical control technology ensures the stability of deep surrounding rock and the safety of mining.

2. Study on Evolution Characteristics of Roadway Surrounding Rock Fissure

The surrounding rock fissures are a kind of structural plane in rock mass and exist in large numbers. It is mainly included in primary and secondary fissures. In actual engineering, rock mass with primary fractures is called fractured rock mass when the external force of rock mass is not enough to destroy its strength. Firstly, the primary cracks in the surrounding rock expand, and then secondary cracks appear. At the same time, the primary cracks of the surrounding rock expand, and the secondary cracks continue to develop. The cracks have different opening degrees, and the cracks are connected to each other, which leads to the failure and instability of the rock mass. The surrounding rock fissures of deep roadways are mainly subjected to the action of pressure, and the secondary fissure is one of the common ones, with uneven distribution and large size differences.

Based on previous research results [24], before deep roadway excavation, rock mass is in a three-way stress environment. Once the roadway is excavated, the three-way stress environment of the rock mass will be destroyed. After the excavation and unloading of roadway, the stress field appears to be redistributed. When the stress of the surrounding rock is higher than the elastic limit and enters the plastic state, the surrounding rock will successively appear in the broken zone, the plastic zone, and the elastic zone from the outside. The three corresponding stress–strain stages in the three zones of the surrounding rock are shown in Figure 1. In the figure, R_0 is tunnel radius, m, R_f is plastic zone radius, m, and R_1 is crushing zone radius, m. In the total stress–strain curve of rock, σ_s is yield strength, MPa, σ_{pk} is peak strength, MPa, and σ_c is residual strength, MPa.

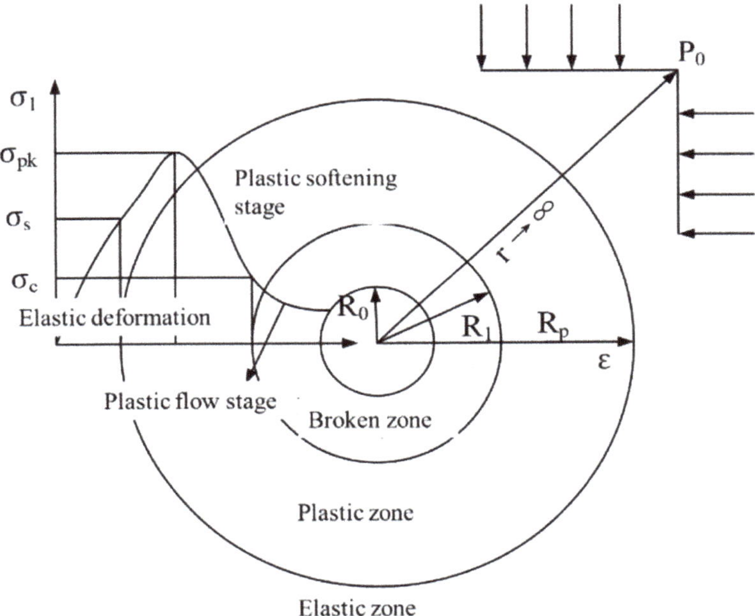

Figure 1. Surrounding rock zoning and rock stress–strain curve model of circular roadway.

After tunnel excavation, stress on the surrounding rock is redistributed, resulting in unidirectional, bidirectional, and tridirectional states of the surrounding rock from the inside to the outside. At the same time, due to the different degrees of damage caused by the rock mass during the stress process, the rock mass is successively located in the fracture zone, the plastic zone, and the original rock stress zone along the radial direction of the roadway. Corresponding to the stress reduction zone, the stress increase zone, and the proto-rock zone, a relatively obvious hierarchical gradient failure pattern is presented. The gradient failure structure of the surrounding rock is shown in Figure 2. According to the permeability characteristics of each zone [25], it can be divided into complete seepage zone, seepage shielding zone, and proto-rock seepage zone. The fractured zone of the roadway is at the stage of the residual strength of the surrounding rock, and the cracks are large and fully developed, forming a complete seepage zone. The axial and circumaxial permeability of the roadway is equivalent, and the grout can be diffused in all directions. When the roadway enters the plastic zone from the outside to the inside, the permeability of the surrounding rock decreases under the action of the structural plane and it cracks, and the grout is mostly diffused with large cracks. This area is called the directional seepage zone. As the radial depth continues to increase, due to the strengthening of the confining pressure, the cracks are subjected to a compression effect, forming fully developed micro and meso cracks. With the increase of depth, the permeability decreases, and the minimum value is found at the critical point of the plastic and elastic zones, which seriously affects the flow of the grout and forms a seepage shield zone. The permeability of the elastic zone is restored to the state of the original rock, namely the original rock seepage zone. The radial seepage zone of the roadway and its permeability characteristics are finally formed, as shown in Figure 3.

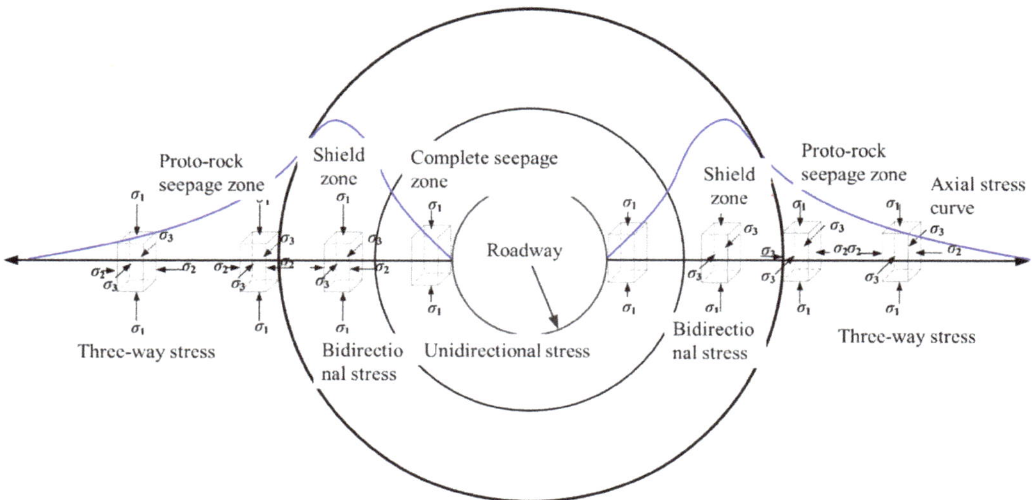

Figure 2. Schematic diagram of gradient failure structure of surrounding rock.

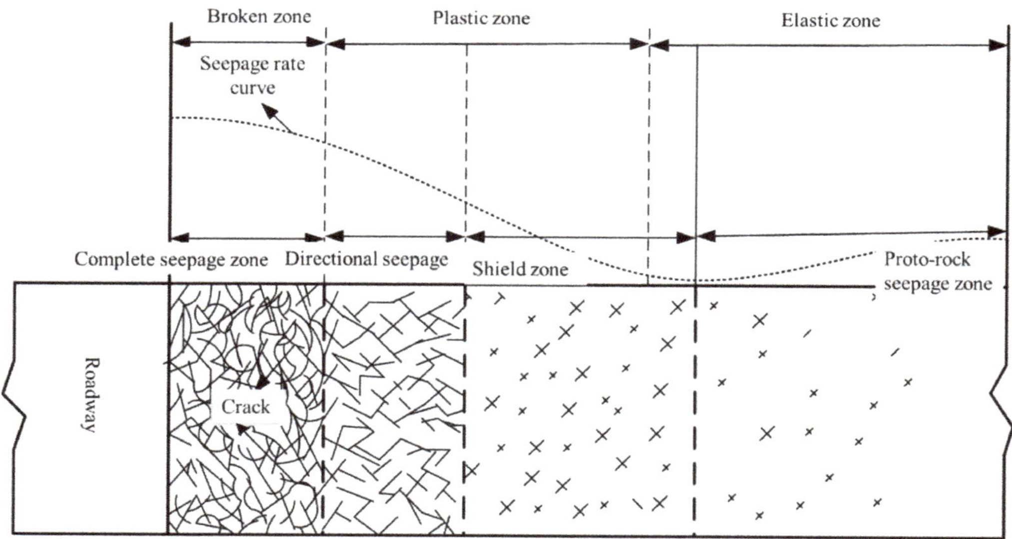

Figure 3. Radial seepage development characteristics of surrounding rocks.

3. Study on Graded Control Mechanism of Gradient Support

3.1. Graded Control Mechanism of Gradient Support

Due to the complexity of deep roadway, there are still a large number of unsolvable scientific problems in roadway management. Traditional support means only combine a variety of support methods, lack the research on the control mechanism of roadways, and do not consider the influence of gradient failure, let alone put forward the control method combined with gradient failure. Therefore, it is difficult to achieve the ideal support control effect. In view of this, combined with the grouting materials developed in this paper, the grouting reinforcement technology is used to put forward a graded control method for gradient support of the surrounding rock, forming a gradient bearing shell model, and the control mechanism is studied in detail. Graded control of gradient support of the surrounding rock mainly refers to adopting corresponding support measures in the corresponding area of the surrounding rock according to the evolution characteristics of the stress gradient of the surrounding rock of the roadway in order to form corresponding gradient support measures in the control area and realize graded control in different areas. For example, the stress gradient in the fracture zone is significant, and high-strength reinforcement materials and bolt support are adopted. While the stress gradient in the plastic zone is weakened, the gradient control is formed step-by-step by adopting the reinforcement material of nanoscale grouting and the anchor cable support.

3.2. Key Ideas of Graded Control for Gradient Support

After the excavation of deep roadway, the stress state will change and a stress gradient field will appear in the radial stress direction, resulting in different damage of the surrounding rock from the inside to the outside and finally making the surrounding rock present a typical stress gradient failure mode. When a roadway is excavated, stress concentration will occur, and gradient failure will be more obvious. In particular, when the stress of the deep roadway surrounding rock is greater than the bearing capacity of the rock, it will eventually lead to zoned failure of the surrounding rock, that is, the fracture zone, the plastic zone, and the elastic zone, as shown in Figure 4a. The uniaxial compressive strength of rock, σ_c, and its variation trend shows the characteristics of increasing zoning, with the lowest strength in the crushing zone and the highest strength in the elastic zone, that is, $\sigma_{c1} < \sigma_{c2} < \sigma_{c3}$. Detailed analysis can be seen in Figure 4b. Element 1 is in the state of primary rock stress before excavation. Once the equilibrium of one side is damaged, the horizontal stress of σ_{21} will be relieved, and the element will change from three-way stress to two-way stress, or unidirectional stress, and its strength will decrease significantly. At the same time, the stress of the original rock is redistributed, showing stress concentration. If the stress of element 1 (assuming that it is in the crushing zone) is greater than its bearing capacity, the element will be destroyed, and the stress will continue to transfer to the interior of the roadway.

Similarly, the horizontal stress value of element 2 (assuming it is in the plastic zone) is small, and the strength of element 2 is greater than that of element 1, while the failure of element 2 continues, and the stress continues to transfer. During the whole stress process, with the stress transfer, the degree of stress concentration decreases, and the strength of the element gradually increases until the stress of the element n (assuming it is in the elastic zone) and subsequent surrounding rock is less than its bearing capacity. Thus, the element is not damaged, and the original stress condition is maintained. The above analysis shows that, in the whole process of stress transfer, the surrounding rock is subject to similar changes in each direction, and then the surrounding rock of the roadway presents different structural levels of failure.

The bearing capacity of rocks in the broken zone and the plastic zone obviously decreases. If no reasonable means of control is taken, the roadway will continue to deform and fail until it becomes unstable. In order to solve the problem of the surrounding rock deformation, the research group, combined with the gradient failure characteristics of the surrounding rock, proposed a graded gradient support control method. In other words, gradient graded support is adopted to realize the integration of the surrounding rock and provides support in the broken zone and the plastic zone. Therefore, the uniformity of the load is controlled by layers to ensure that the strength of the surrounding rock in the broken zone and the plastic zone is restored to the initial stress strength or close to the state of the surrounding rock in the elastic zone, that is, the implementation of the original $\sigma_{c1} < \sigma_{c2} < \sigma_{c3}$ transforms into $\sigma^{*}_{c1} \approx \sigma^{*}_{c2} \approx \sigma^{*}_{c3}$.

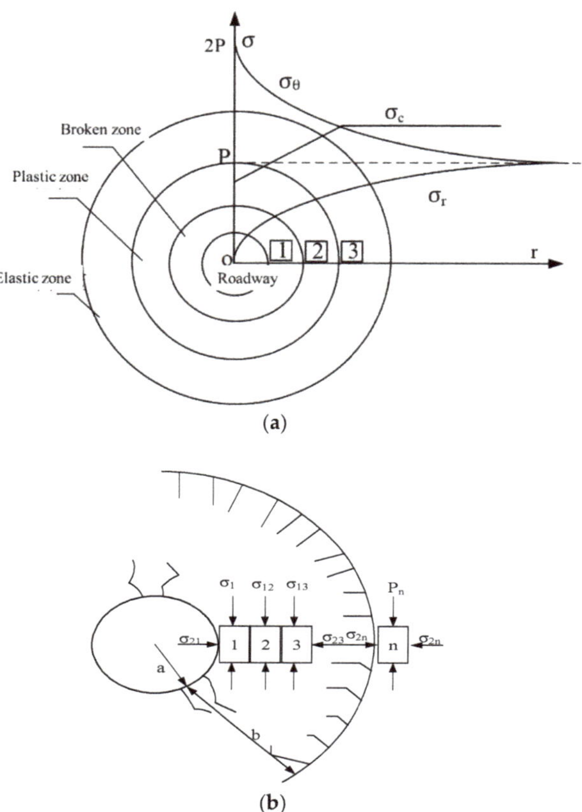

Figure 4. Mechanical model after roadway excavation; (**a**) Mechanical model of surrounding rock; (**b**) Stratified failure of surrounding rock physical state [26].

3.3. Hierarchical Control Principle Model of Gradient Support

The changes of the stress gradient cause the surrounding rock to present unidirectional, bidirectional, and tridirectional stress conditions from the inside to the outside. At the same time, due to the different degrees of damage caused by the rock mass during the stress process, the fracture zone, plastic zone (fracture zone), and elastic zones are formed, showing relatively significant hierarchical gradient failure. The gradient failure leads to a significant reduction in the bearing capacity of the rock mass. If the rock mass is not treated in time under continuous stress, large deformation or even instability of roadway surrounding rock will occur. Therefore, according to the gradient failure, this paper puts forward the graded control method of gradient support to further enhance or even restore the bearing capacity of the rock mass. As shown in Figure 5, curve I represents the change rule of rock mass strength caused by gradient failure, while curve II represents the ideal change of the rock mass strength after graded control with gradient support. The gradient bearing shell outside the roadway is realized by means of bolting and high-strength grouting. Since the cracks and pore sizes in the plastic zone gradually decrease along the radial stress direction of the roadway, the open cracks also gradually transform into closed cracks. It is difficult for ordinary grouting materials to complete good consolidation and filling, and they still cannot prevent the surrounding rock from developing from the plastic zone to the broken zone. Therefore, in the plastic zone (fractured zone), nano-base grouting materials are used to achieve graded reinforcement, and the internal three-dimensional gradient bearing shell is formed by combining with the anchor cable in order to achieve the transformation from curve I to curve II. The gradient bearing shell includes the broken zone and the plastic zone, which is the gradient support technology corresponding to the stress gradient change established by the pointer and forms the supporting shell with corresponding bearing capacity. We collectively call it the gradient bearing shell [27]. The shell properties formed in the broken zone and the plastic zone are different, in order to distinguish the gradient pressure shell (broken zone) and the gradient pressure shell (plastic zone).

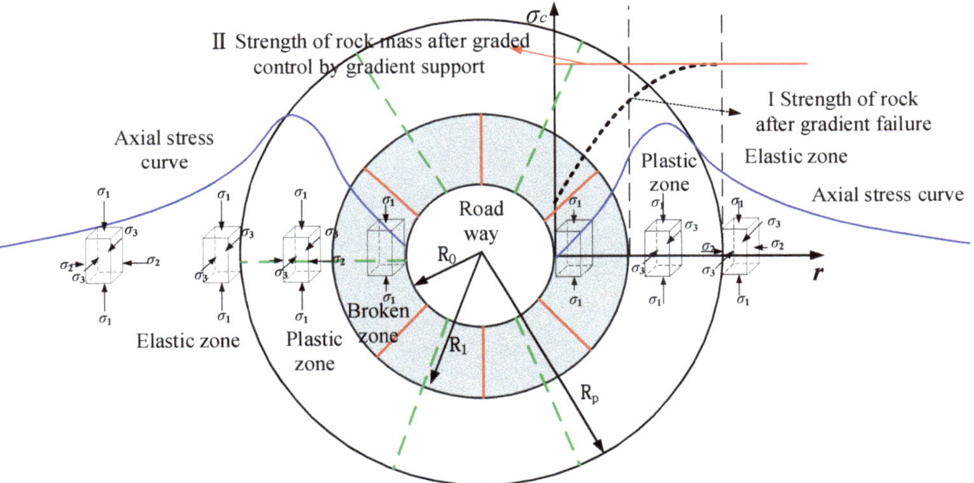

Figure 5. Mechanical model of graded support control.

4. Discussion on the Injectivity of Reinforcement Materials

4.1. Traditional Reinforcement Material

The stress transition caused by roadway excavation leads to the continuous development of surrounding rock cracks, which reduces the strength of the surrounding rock, causing large deformation that seriously affects its stability. The surrounding rock reinforcement can seal the fracture to a certain extent, improve its mechanical properties, and ensure the stable deformation. However, at present, Portland cement is the main material for roadway grouting. However, due to its large particle size, late consolidation cracking, and low grout adhesion, the performance of Portland cement material is greatly weakened in the underground environment with rich water and ion erosion. Combined with the seepage development characteristics of the surrounding rock, it can be seen that the large particle size of traditional materials causes the poor diffusion of grout and the existence of many deep micro-cracks, which leads to the weakening of grout diffusion radius and even causes problems, such as the non-injection of cracks. At the same time, the high grouting pressure method adopted for micro-cracks easily causes the fracture to coalesce, increases the leakage channels, affects the slurry diffusion, and greatly weakens the reinforcement effect. However, the traditional grouting mostly adopts the one-time reinforcement technology, which eventually leads to poor grouting ability and other problems, as shown in Figure 6.

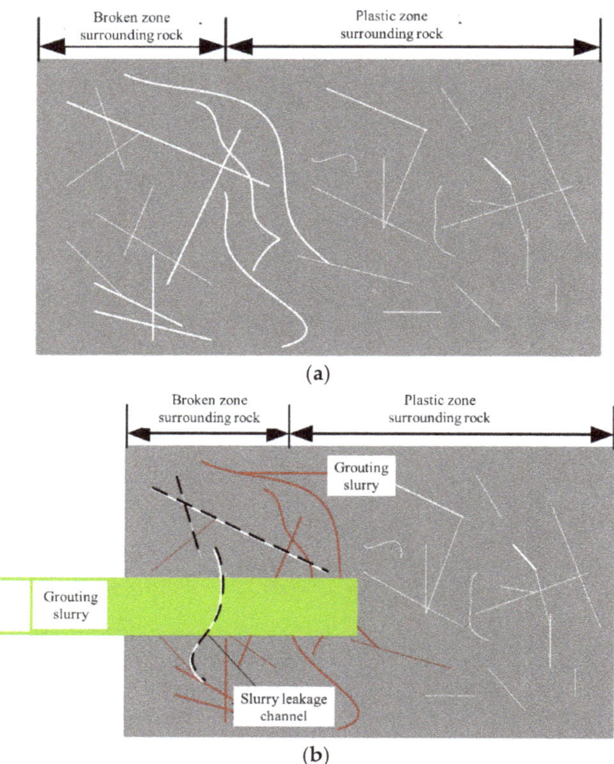

Figure 6. Chematic diagram of traditional grouting [28]. (**a**) Without grouting in surrounding rock; (**b**) traditional grouting in surrounding rock.

Regarding the injectivity of grout of grouting material, according to the research work completed by scholars, the injectivity standard of grout can be summarized as follows: The crack opening is D_r, and grout particle size d must meet the formula [29] (1):

$$\frac{D_r}{d} > 3 \tag{1}$$

Assuming that the grout can penetrate into the cracks of the injected surrounding rock, the crack opening should meet the following requirements:

$$D_r > 3d > 0.18 \text{ mm} \tag{2}$$

The crack opening of the surrounding rock must meet the conditions of Formula (2); the slurry can be permeated into the filled fissure. However, the change of underground stress leads to the evolution of structure hierarchy of the surrounding rock, which is manifested as obvious crack changes from the outside to the inside of the roadway. That is, the cracks close to the roadway are dense and wide, and sparse and narrow cracks continue to appear inward [30]. This phenomenon shows that not all cracks meet this condition, and micro-cracks (plastic zone cracks) have difficulty with meeting the above requirements, which will inevitably increase the difficulty of grout diffusion, reduce the stability of the surrounding rock after reinforcement, and cause continuous deformation.

The superfine cement material improves the injectivity of the slurry to some extent. Due to the gradient stress, there are many cracks in the plastic zone, and the crack opening is very small, even less than 20 μm, resulting in grout that cannot be injected into the crack. The organic grouting material has good injectivity and high adhesion to the surrounding rock. However, its price is too high, and it pollutes the water, corrodes the human body, and is not natural enough. Therefore, many mines list organic materials as contraband, such as polyurethane and Malisan.

For this reason, the surrounding rock of deep mine reinforcement requires materials with high permeability and high viscosity. High permeability can be achieved by changing the particle size of the material, and high viscosity can be achieved by changing the chemical reaction between the crack and the slurry. When the crack opening is 3 times of the particle size of the reinforcement material, the grout can be injected into the crack. If the difference is greater than 3 times, the grout cannot be injected into the crack and will gradually show a bridging state at the crack mouth. At the same time, considering the complexity of the deep mine environment, it is highly likely to face water-rich and ionic erosion. It is difficult for traditional grouting materials to resist ion erosion, which is bound to cause this. Moreover, even after grouting reinforcement, its performance will be greatly reduced, and it cannot meet the requirements of the surrounding rock control. Therefore, it is necessary to further develop reinforcement materials resistant to ion erosion on the basis of satisfying the filling.

4.2. Graded Control Technology of Gradient Support

As can be seen from Figure 7, radial stress σ_r and tangential stress σ_θ of the surrounding rock are greatly affected after tunnel excavation. The radial stress increased gradually with the increase of distance and was almost zero near the edge of the surrounding rock. However, the tangential stress gradually decreased with the increase of distance and reached its maximum near the edge of the surrounding rock. However, both of them are close to stable at the distance of excavation and are hardly affected by excavation. At the distance of r_0–$2r_0$, there was a large difference in stress evolution. The slope of the straight line in the dotted box indicates that the stress of the surrounding rock gradually increased under the influence of the buried depth [30]. The surrounding rock of the deep mining roadway is greatly affected by secondary stress, and the stress gradient field that formed gradually became prominent. There are obvious high stress gradients around deep roadways, which is the main factor leading to the continuous failure of roadways.

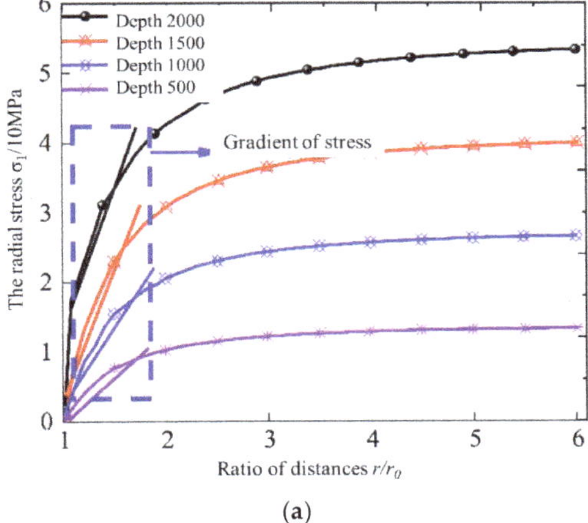

(a)

Figure 7. *Cont.*

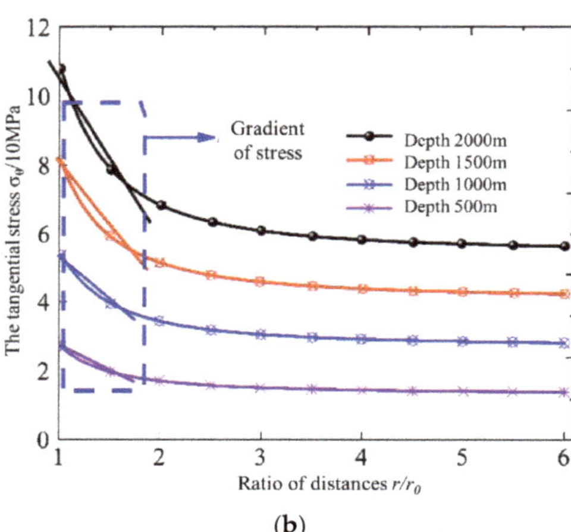

(b)

Figure 7. Distribution diagram of surrounding rock stress after roadway excavation. (**a**) The radial stress; (**b**) The tangential stress.

At the same time, considering the evolution of the surrounding rock structure caused by gradient failure, the concrete manifestation is that the cracks close to the roadway are dense and wide, and the cracks continue to be sparse and narrow. The traditional grouting materials can only inject cracks with a large crack opening, and micro-cracks cannot reach the injection range. However, for the crack distribution caused by gradient failure of the surrounding rock, the crack opening is small, and there are many cracks in the plastic zone. If the reinforcement cannot be completed, the surrounding rock is bound to be damaged or even unstable. Therefore, this paper studied the small particle grouting reinforcement material, combined with high-strength cement material, and proposed a gradient support classification control method, hoping to improve or even restore the bearing performance of the surrounding rock, that is, the broken area is reinforced by bolt and high-strength grouting. The cracks and pore sizes in the plastic zone gradually become smaller along the radial stress direction of the roadway, and the open cracks also gradually transform into closed cracks. The graded reinforcement is achieved by using nano-based grouting materials, and the internal three-dimensional gradient pressure shell is formed by combining with the anchor cable. After the reinforcement of small particle size grout, all the cracks in the surrounding rock can be filled using multi-scale grading to meet the stability control problem of the surrounding rock. The research work should be further combined with the quantitative characteristics of deep rock cracks, establish a quantitative model, and carry out field tests in other mines in order to make it more suitable for deep, complex geological conditions.

5. Conclusions

According to the gradient failure characteristics of deep roadway surrounding rock, the main conclusions obtained in this paper are as follows:

(1) After roadway excavation, the stress of the deep surrounding rock is redistributed. Along the radial direction of the roadway, it is divided into stress reduction zone, stress increase zone, and original rock stress zone. According to the permeability characteristics, it can be divided into complete seepage zone, seepage shield zone, and proto-rock seepage zone.

(2) When the crack opening degree of the surrounding rock meets certain conditions, slurries can permeate into the crack. However, the deep surrounding rock cracks are dense and wide near the roadway, while the deep ones are sparse and narrow. Changing the particle size of the material can improve the grouting ability. If the

difference between the particle size of the grouting material and the crack is greater than 3 times, the grouting cannot be injected into the crack, and the bridge state will gradually appear at the crack mouth, greatly reducing the reinforcement effect.

(3) The equivalent graded gradient support is proposed, and the mechanical model of gradient support is established. The gradient graded support is adopted to achieve the integration of the surrounding rock and provide support to the broken zone and to the plastic zone. Stratified control of load uniformity ensures that the strength of the surrounding rock in the broken zone and the plastic zone is restored to the initial stress strength or close to the state of the surrounding rock in the elastic zone, that is, the implementation of the original $\sigma_{c1} < \sigma_{c2} < \sigma_{c3}$ transforms into $\sigma^*_{c1} \approx \sigma^*_{c2} \approx \sigma^*_{c3}$.

Author Contributions: Conceptualization, Z.H.; methodology, Z.L.; formal analysis, F.D.; investigation, Z.C.; data curation, Z.H.; writing—original draft preparation, Z.H. and Z.L.; writing—review and editing, Z.H.; visualization, C.Z. All authors have read and agreed to the published version of the manuscript.

Funding: This study was supported by the Science and Technology Project of Henan Province (222102320004) by Zijie Hong, Central Plains Talent Program—Central Plains Young Top Talent (Central Plains Young Postdoctoral Innovative talent) by Zijie Hong, State and Local Joint Engineering Laboratory for Gas Drainage and Ground Control of Deep Mines (SJF2209) by Zijie Hong, the opening project of Henan Key Laboratory of Underground Engineering and Disaster Prevention (Henan Polytechnic University) by Zijie Hong, Civil Engineering School of young teachers Research ability Enhancement Fund, Henan Polytechnic University by Zijie Hong; Doctoral Fund of Henan Polytechnic University (B2022-25) by Zijie Hong, and the National Natural Science Foundation of China (52174073) by Zhenhua Li. The authors are grateful to the editor and reviewer for discerning comments on this paper.

Institutional Review Board Statement: The study did not require ethical approval.

Informed Consent Statement: Informed consent was obtained from all subjects involved in the study.

Data Availability Statement: Data will be provided upon request to the corresponding author.

Conflicts of Interest: The authors declare no conflict of interest.

References

1. Bai, E.; Guo, W.; Tan, Y. Negative externalities of high intensity mining and disaster prevention technology in China. *Bull. Eng. Geol. Environ.* **2019**, *78*, 5219–5235. [CrossRef]
2. Huang, Q.; He, W. Research on overburden movement characteristics of large mining height working face in shallow buried thin bedrock. *Energies* **2019**, *12*, 4208. [CrossRef]
3. He, M.C.; Xie, H.P.; Peng, S.P.; Jiang, Y.D. Study on rock mechanics in deep mining engineering. *Chin. J. Rock Mech. Eng.* **2005**, *24*, 2803–2813.
4. Chen, J.; Zhao, J.; Zhang, S.; Zhang, Y.; Yang, F.; Li, M. An experimental and analytical research on the evolution of mining cracks in deep floor rock mass. *Pure Appl. Geophys.* **2020**, *177*, 5325. [CrossRef]
5. Yin, Q.; Wu, J.Y.; Jiang, Z.; Zhu, C.; Su, H.; Jing, H.; Gu, X. Investigating the effect of water quenching cycles on mechanical behaviors for granites after conventional triaxial compression. *Geomech. Geophys. Geo-Energy Geo-Resour.* **2022**, *8*, 77. [CrossRef]
6. Liu, C.; Yang, Z.; Gong, P.; Wang, K.; Zhang, X.; Zhang, J.; Li, Y. Accident analysis in relation to main roof structure when longwall face advances toward a roadway: A case study. *Adv. Civ. Eng.* **2018**, *2018*, 3810315. [CrossRef]
7. Yu, W.; Pan, B.; Zhang, F.; Yao, S.; Liu, F. Deformation characteristics and determination of optimum supporting time of alteration rock mass in deep mine. *KSCE J. Civ. Eng.* **2019**, *23*, 4921–4932. [CrossRef]
8. Ren, D.Z.; Wang, X.Z.; Kou, Z.H.; Wang, S.; Wang, H.; Wang, X.; Tang, Y.; Jiao, Z.; Zhou, D.; Zhang, R. Feasibility evaluation of CO_2 EOR and storage in tight oil reservoirs: A demonstration project in the Ordos Basin. *Fuel* **2023**, *331*, 125652. [CrossRef]
9. Wang, Y.; Zhu, C.; He, M.C.; Wang, X.; Le, H.L. Macro-meso dynamic fracture behaviors of Xinjiang marble exposed to freeze thaw and frequent impact disturbance loads: A lab-scale testing. *Geomech. Geophys. Geo-Energy Geo-Resour.* **2022**, *8*, 154. [CrossRef]
10. He, M.; Sui, Q.; Li, M.; Wang, Z.; Tao, Z. Compensation Excavation Method Control for Large Deformation Disaster of Mountain Soft Rock Tunnel. *Int. J. Min. Sci. Technol.* **2022**, *5*, 951–963. [CrossRef]
11. Alsayed, M.I. Utilising the Hoek triaxial cell for multiaxial testing of hollow rock cylinders. *Int. J. Rock Mech. Min. Sci.* **2002**, *39*, 355–366. [CrossRef]
12. Kulatilake, P.H.S.W.; Park, J.; Malama, B. A new rockmass failure criterion for biaxial loading conditions. *Geotechnol. Geol. Eng.* **2006**, *24*, 871–888. [CrossRef]

13. Unver, B.; Yasitli, N.E. Modelling of strata movement with a special reference to caving mechanism in thick seam coal mining. *Int. J. Coal Geol.* **2006**, *66*, 227–252. [CrossRef]

14. Ju, Y.; Wang, Y.; Su, C.; Zhang, D.; Ren, Z. Numerical analysis of the dynamic evolution of mining-induced stresses and fractures in multilayered rock strata using continuum based discrete element methods. *Int. J. Rock Mech. Min. Sci.* **2019**, *113*, 191–210. [CrossRef]

15. Duan, B.; Xia, H.; Yang, X. Impacts of bench blasting vibration on the stability of the surrounding rock masses of roadways. *Tunnel. Undergr. Space Technol.* **2018**, *71*, 605–622. [CrossRef]

16. Yu, W.; Liu, F. Stability of close chambers surrounding rock in deep and comprehensive control technology. *Adv. Civ. Eng.* **2018**, *2018*, 6275941. [CrossRef]

17. Liu, C. Research on deep-buried roadway surrounding rock stability and control technology. *Geotech. Geol. Eng.* **2018**, *36*, 3871–3878. [CrossRef]

18. Bao, Y.D.; Chen, J.P.; Su, L.J.; Zhang, W.; Zhan, J. A novel numerical approach for rock slide blocking river based on the CEFDEM model: A case study from the Samaoding paleolandslide blocking river event. *Eng. Geol.* **2023**, *312*, 106949. [CrossRef]

19. Kang, H.P.; Wang, G.F.; Jiang, P.F.; Wang, J.C.; Zhang, N.; Jing, H.W.; Huang, B.X.; Yang, B.G.; Guan, X.M.; Wang, Z.G. Conception for strata control and intelligent mining technology in deep coal mines with depth more than 1000 m. *J. China Coal Soc.* **2018**, *43*, 1789–1800.

20. Xie, H.; Peng, S.; He, M. *Basic Theory and Engineering Practice in Deep Mining*; Science Press: Beijing, China, 2005; pp. 1–35.

21. Nikbakhtan, B.; Osanloo, M. Effect of grout pressure and grout flow on soil physical and mechanical properties injet grouting operations. *Int. J. Rock Mech. Min. Sci.* **2009**, *46*, 498–505. [CrossRef]

22. Hong, Z.; Zuo, J.; Zhang, Z.; Liu, C.; Liu, L.; Liu, H.Y. Effects of nano-clay on the mechanical and microstructural properties of cement based grouting material in sodium chloride solution. *Constr. Build. Mater.* **2020**, *245*, 118420. [CrossRef]

23. Varol, A.; Dalgic, S. Grouting applications in the Istanbul metro, Turkey. *Tunnel. Undergr. Space Technol.* **2006**, *21*, 602–612. [CrossRef]

24. Wang, H.; Li, H.; Tang, L.; Ren, X.; Meng, Q.; Zhu, C. Fracture of two three-dimensional parallel internal cracks in brittle solid under ultrasonic fracturing. *J. Rock Mech. Geotech. Eng.* **2022**, *14*, 757–769. [CrossRef]

25. Zhang, N.; Xu, X.L.; Li, G.C. Fissuere-evolving laws of surrounding rock mass of roadway and control of seepage disasters. *Chin. J. Rock Mech. Eng.* **2009**, *28*, 331–335. [CrossRef]

26. Dong, F. *Support Theory and Application Technology of Roadway Surrounding Rock Loose Zone*; China Coal Industry Press: Beijing, China, 2001; pp. 20–45.

27. Zuo, J.P.; Hong, Z.J.; Yu, M.L.; Liu, H.Y.; Wang, Z.B. Research on gradient support model and classification control of broken surrounding rock. *J. China Univ. Min. Technol.* **2022**, *51*, 221–231.

28. Zhang, Z. Study on the Mechanism of Grouting Reinforcement of Weak Medium Filling Cracks. Ph.D. Dissertation, Shangdong University, Jinan, China, 2015.

29. Guo, Y.; Zhang, B.L.; Zheng, X.G.; Zhou, W.; Wei, X.L.; Zhu, D.X. Grouting reinforcement technology of superfine cement with wind oxidation zone based on time-varying rheological parameters of slurries. *J. Min. Saf. Eng.* **2019**, *36*, 339–344.

30. Zuo, J.P.; Wei, X.; Wang, J.; Liu, D.J.; Cui, F. Investigation of failure mechanism and model for rocks in deep roadway under stress gradient effect. *J. China Univ. Min. Technol.* **2018**, *47*, 478–485.

MDPI
St. Alban-Anlage 66
4052 Basel
Switzerland
Tel. +41 61 683 77 34
Fax +41 61 302 89 18
www.mdpi.com

Energies Editorial Office
E-mail: energies@mdpi.com
www.mdpi.com/journal/energies